Discover Entropy
and the
Second Law of Thermodynamics

A Playful Way of Discovering a Law of Nature

Discover Entropy
and the
Second Law of Thermodynamics

**A Playful Way of
Discovering a Law of Nature**

Arieh Ben-Naim

The Hebrew University of Jerusalem, Israel

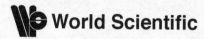 **World Scientific**

NEW JERSEY · LONDON · SINGAPORE · BEIJING · SHANGHAI · HONG KONG · TAIPEI · CHENNAI

Published by

World Scientific Publishing Co. Pte. Ltd.

5 Toh Tuck Link, Singapore 596224

USA office: 27 Warren Street, Suite 401-402, Hackensack, NJ 07601

UK office: 57 Shelton Street, Covent Garden, London WC2H 9HE

Library of Congress Cataloging-in-Publication Data
Ben-Naim, Arieh.
    Discover entropy and the second law of thermodynamics : a playful way of
discovering a law of nature / Arieh Ben-Naim.
        p. cm.
    Includes bibliographical references and index.
    ISBN-13: 978-981-4299-75-6 (hardcover : alk. paper)
    ISBN-10: 981-4299-75-8 (hardcover : alk. paper)
    ISBN-13: 978-981-4299-76-3 (pbk. : alk. paper)
    ISBN-10: 981-4299-76-6 (pbk. : alk. paper)
    1. Entropy. 2. Second law of thermodynamics. 3. Thermodynamics. I. Title.
    QC318.E57B458 2010
    536'.73--dc22
                                        2010023319

British Library Cataloguing-in-Publication Data
A catalogue record for this book is available from the British Library.

Typeset by Stallion Press
Email: enquiries@stallionpress.com

Printed in Singapore.

Publishers' page

This book is dedicated to
Claude E. Shannon

Painting by Alexander Vaisman

Low Entropy Room

High Entropy Room

Contents

Preface

This book is a sequel to my previous book, *Entropy Demystified*, which was published in 2007. While writing that book, I had in mind the layperson as my target audience. I presumed that the potential reader need not have any background in science and all that was required to understand that book was plain common sense and a willingness to use it.

From the feedback I got from both scientists and non-scientists, it became clear to me that I was wrong. One reader of the book, who is not a scientist but has a PhD in musicology, told me it was too difficult to understand. I asked her why. After all, I wrote it for the layperson. Her answer was: "You are right, the book *is* written for the layperson, but I am not even a layperson".

With this and similar comments I have received, I decided that I must bring down the level of the book even further. In a sense, this book could be viewed as a prequel to my previous book. I have done my utmost to translate relatively difficult and highly mathematical concepts into a simple, clear, and familiar language. I hope I have succeeded in doing that.

This book was planned quite differently from the previous one. It is addressed to anyone who is curious about the world around them and is willing either to perform or to follow some simple experiments with marbles distributed in boxes or cells. There is no need to know any mathematics to be able to understand the contents of this book.

In Chapter 2 of my previous book, I discussed in detail two of the central concepts underpinning the Second Law of Thermodynamics: probability and information.

In the present book, however, I refrain from doing that. Instead, I have based the entire book on what you, the reader, already know of these two central concepts.

Many who have read my previous book have commented that there are too many footnotes in the text, which distract the attention of the reader. I have therefore done away with footnotes in this book and have instead moved them to a separate chapter (Chapter 8), where some technical issues are discussed very briefly. Now the reader can read and hopefully understand the book without any interruptions.

Notwithstanding the lack of distracting notes, I have deliberately strewn within the pages of this book some distractive materials that I refer to as "snacks". Some of these "snacks" are simple, entertaining stories, some others are "brainteasers," but all are meant to ease your way through the book and make the reading experience more enjoyable.

Understanding a scientific book, even at the most elementary level, requires *active reading*. This is especially true when you are expected to follow experiments, analyze results, search for regularities or irregularities, recognize patterns that are common to all the experiments, and hopefully discover something new.

Active reading may also be beneficial to your health. As you might know, until recently it was believed that we are born with a fixed number of neurons in our brains. As we age, many neurons fade out and die, resulting in the reduction of brain function. However, it has now been discovered that new neurons are born even after childhood. These new neurons are short-lived and, unless you use them, you lose them!

It is therefore to your advantage and benefit to read the book *actively*, as this will maintain your mental capabilities. To work through a chapter a day might well keep Dr Alzheimer away!

There is another health-related benefit of these entertaining "snacks." It is a well documented fact that a smile or a laugh can significantly boost your immune system: I hope some of these "snacks" will elicit a healthy smile.

An effective way to read this book actively is to take a break, have a "snack", and try to repeat an argument or re-derive a conclusion by yourself. It is not uncommon for a student to believe that he or she understands an argument when in fact there is only a delusion of understanding. In my experience, I have found that the best criterion I can apply to test my understanding of what I have read is to explain it to someone else. Entropy and the Second Law are quite tricky to understand. If you only scan along the lines of the book you might fall into the trap of having a false sense of understanding what you have read. For this reason, you should be engaged in active reading and test your understanding by trying to explain what you have read to someone else.

The structure of this book is as follows.

Chapter 1 provides a very brief introduction to the Second Law of Thermodynamics. This chapter is not meant to *teach* you the Second Law. You will have to discover it yourself.

Chapter 2 presents two central concepts: probability and information. Again, this chapter is not meant to *teach* you either probability or information. Instead, it is meant to show that you already know what you need to know for understanding the Second Law. I hope that this chapter will convince you that you can understand the entire book.

In Section 2.2, we shall see that children as young as 6–9 years old already have an intuitive understanding of the concept of probability, even though they cannot define this concept.

Section 2.3 is devoted to the concept of information and its measure. I presume that everyone knows what "information" is, although it is very difficult to define. However, in this chapter we shall not discuss "information" itself but some measure of the *size* of information.

In fact, we shall zero-in on a specific, more limited, and more precise concept of Shannon's measure of information (*SMI*), which can also be grasped easily by children who have played the popular "20-questions" (20Q) game. At quite an early stage, children intuitively feel that certain strategies of asking questions are more efficient than others; that is, efficient in the sense of asking the minimum number of questions to obtain the required information. It is here that Shannon's measure of information is introduced, not mathematically and not formally, but indirectly through the more familiar 20Q game.

Within Chapters 3–5 lies the core of the book. Here, we shall carry out various experiments with marbles in cells. The experiments were designed in such a way that you will not get bogged down in numerous, fruitless trial and error cases.

Although I am not going to teach you the Second Law, I will help you with the methodology of the scientific work. I presume you have heard of Charles Darwin. During his long and extensive trips around the world, Darwin *collected* and *registered* a huge number of lifeless and seemingly unrelated items. Those items would have remained lifeless forever had Darwin not penetrated into the significance of these items and discovered how the diversity of living creatures has emerged from the lifeless and seemingly unrelated items.

While learning the methodology of the scientific work, we shall make some important discoveries. In Chapter 3, we shall discover the *uniform distribution* and how it is related to the Second Law. In Chapters 4 and 5, we shall discover the Boltzmann distribution and the Maxwell–Boltzmann distribution, and how they are related to the Second Law.

In Chapter 6, we shall summarize all that we have learned from the experiments described in Chapters 3–5. We shall formulate the Second Law for the system of marbles in cells. In a nutshell, we shall first define the "size" of the games played in Chapters 3–5. This "size" measures how easy or difficult the game is. This measure will be the analog of the entropy. The Second Law, in game language, states: Every game will always evolve in such a way as to become more difficult to play. This evolution will stop when we reach the equilibrium state, at which point the game becomes most difficult to play.

I believe that these analogs of entropy and the Second Law will be easy to grasp in the world of marbles distributed in boxes or cells. Once you feel comfortable with these concepts in the context of the marbles, there remains only the technical chore of translating from the language of the marbles in cells into the language of real particles in real boxes. As a result of this translation, the concepts of entropy and the Second Law in will be ripe for the picking.

Chapter 7 is a translation of Chapter 6, from the language of marbles in cells to the language of real systems of particles in boxes. In this chapter we shall discuss the three central questions relevant to the Second Law: What is entropy? Why does it change in one direction only? How does the system evolve from any arbitrary initial state to the eventual equilibrium state? You will also learn the workings of the Second Law in phenomena that you encounter in your day-to-day life.

You certainly know that if you spray even just a small amount of perfume in the corner of a room, you will notice that after a few minutes it will expand and occupy the entire room. But did you also know that the spread of the perfume molecules in the entire room will be *uniform*? This is not a trivial matter to prove, as we shall discover in Chapter 3. What about the distribution of the same drop of perfume in a column of air that extends vertically towards the sky? This is not a trivial question either. The distribution will be far from uniform. Thanks to Boltzmann, pilots could measure their altitude *without* sending any signal to the earth. If you cannot appreciate this achievement, think about measuring the distance between yourself and any object at some distance from you. Can you measure the distance without sending something to that object? You cannot. You will either have to go there, send someone there, or send a beam of light and measure the time it takes to come back.

Returning to the same drop of perfume, did you ever ask yourself *how* the molecules managed to fill the entire space, considering that they were initially concentrated in one corner of the room? You might have heard that molecules are in constant motion and that this motion "drives" the molecule to reach every accessible point in the room. You might even know that when the temperature is higher, the spreading of the perfume throughout the room will be faster. But do you know the precise distribution of velocities? We shall also become familiar with this distribution in Chapter 5.

Science has evolved in myriads of ways. Scientists use whatever tools are available, either theoretical or experimental, to measure or calculate predetermined quantities. Others may stumble across a new problem and perhaps hit upon a new discovery serendipitously. In most cases, however, a great amount of tedious and routine work, either through experimentation or calculation, is involved before one discovers a new pattern of interest.

In this book I shall try to give you the feeling of, or perhaps the delusion of, discovering a law of nature. In a sense you will use the trial and error method to discover entropy and the Second Law, using a set of experiments that I have designed in order to lead you to that goal along a relatively short path.

Here is my promise to you: If you read the book actively and critically — please send me any errors you find — you will not only boost your immune system, not only keep Dr Alzheimer away, not only *understand* the meaning of entropy and why it behaves in such a peculiar way, but you will also learn about three of the most important probability distributions in science. You will see how these distributions spawned from Shannon's three theorems. These theorems are highly mathematical, yet you will discover them through experimentation with simple games. And above all, you will experience the joy of discovering one of the most mysterious concepts in physics.

This book is dedicated to Claude Shannon. It is well known that Shannon did not contribute anything to thermodynamics or to statistical mechanics. However, this book is not *about* thermodynamics but about understanding entropy and the Second Law. Shannon's measure of information (which we shall use throughout the book) has provided us with a beautiful and a powerful tool to penetrate into the origin of various molecular distributions underpinning the concept of entropy and of the Second Law of Thermodynamics. Shannon's measure of information has transformed entropy from a vague, confusing, and very often mysterious quantity, into a precise, objective, familiar, and crystal-clear concept. Although it was not Shannon's intention to create a tool for understanding thermodynamics, I feel it is fit to pay tribute to Shannon's extraordinary tool, which has been found so useful in so many diverse fields of research.

If you are pondering on the question of whether you can read and understand this book, let me suggest a simple test. On the next page you will find a self-testing kit. It is designed to check your competence in understanding the contents of this book. The first checks your sense of probabilities. The second will tell you if you know how to group objects into "groups of equivalence," and if you are smart enough to play the 20Q game successfully.

I am confident that if you can *read* this test, then you can also pass the test, and are therefore able to read this book.

Arieh Ben-Naim
Department of Physical Chemistry
The Hebrew University of Jerusalem
Jerusalem, Israel 91904
Email: arieh@fh.huji.ac.il
URL: www.ariehbennaim.com

Self-Testing Kit

Probability

You are presented with two urns. Each contains different numbers of blue and red marbles. You know exactly how many blue and red marbles are in each urn. You have to choose an urn and then pick one marble from it blindfolded. If you pick a blue marble you get a prize, but you get nothing if it is a red marble. Which urn will you choose to pick the marble from?

1. The easy problem:

   Left urn: 4 blue and 4 red marbles

   Right urn: 8 blue and 6 red marbles

2. The difficult problem:

   Left urn: 4 blue and 4 red marbles

   Right urn: 8 blue and 12 red marbles

   Question: Why did I label the two choices as "easy" and "difficult"?

   See answer in Note 1.

Measure of Information

Look at the two figures below. Each contains 16 different marbles — some are pictures of real marbles while others were drawn using the computer program Mathematica™. I am thinking of one specific marble. You have to find out which marble I am thinking of. You are allowed to ask only questions that can be answered with YES or NO.

You have to pay 1 US$ for each answer you receive. When you find the marble I was thinking of, you will get 8 US$. How would you plan to ask questions? Presuming you want to maximize your earnings, can you estimate how much you will gain, or lose, if you play this game 100 times?

Easy: First game.

Difficult: Second game.

Questions:

1. Why did I label the two games "easy" and "difficult"?
2. Do you think you can guarantee your gaining in this game?
3. Suppose that you are not allowed to ask questions about the locations of the marbles (e.g. "Is the marble in first row?"): How would you plan to ask questions?

See answers in Note 2.

Measure of Information

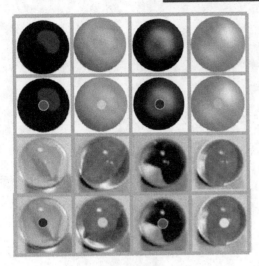

Easy Problem

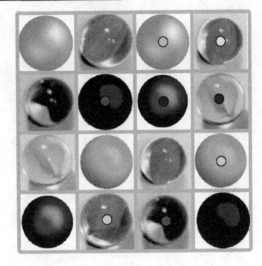

Difficult Problem

END OF TEST

If you feel you know the answers to these questions, you have passed the test.
See also Notes to Self-Testing Kit.

List of Abbreviations

E	Energy
S	Entropy
W	Multiplicity
NB	Number of boxes
NC	Number of cells
NM	Number of marbles
$(NM;NC)$	A game of NM marbles in NC cells
PR	Probability
20Q	20 questions
TA	Total area (of cloth)
TL	Total length (of strings)
SMI	Shannon's measure of information
UDG	Uniformly distributed game
$NUDG$	Non-uniformly distributed game
$ANOQONTAITSS$	Average number of questions one needs to ask in the smartest Strategy

Comment: You do not have to memorize those abbreviations, as I shall repeat their meaning many times throughout the book.

Acknowledgement

The list of colleagues, friends and other persons who contributed to this book is quite long and I might easily miss someone to whom I owe a debt of gratitude. Therefore, if you are reading this book and you see that I failed to mention your contribution or help, please send me an email and I will add your name in the next printing.

I would like to thank first and foremost Alexander Vaisman for graciously offering to produce the wonderful drawings strewn within the pages of this book. I admire his ability to grasp my often sketchy instructions, and with his deft and artful strokes to translate them into beautiful drawings that convey clear messages.

Most of the illustrations were created with the Mathematica™ program. I want to thank Wolfram Research for producing such powerful computation and graphics tools.

Many have sent me either their own articles or referred me to the literature, where I have read fascinating reports on research regarding children's perception of probability, as well as children playing the 20Q game.

Special thanks are also due to Ruma Falk and Marc Marschark, Robert Siegler, and Anne Schlottmann. I am also grateful to Russ Abbott, Ron Barry, Franco Battaglia, Laura Burns, Peter Freund, Zack Galler, Mark Goretsky, Paul King, John Knight, Harvey Leff, Michael Lewis, Robert Mazo, Andrew Smith, Eric Weinberger, Samuele Zampini and Seth Zimmerman, who took time out to read parts or the whole book and offered useful comments.

While working on this book, I have enjoyed collaborating with the staff of World Scientific Publishers. I want to thank the editor Ms Sook-Cheng Lim for her friendly and gracious assistance throughout the period of over five year since I started to work with World Scientific. I am also indebted to the proofreader Sri Varalaxmi, the typesetter Stallion Press as well as to the artist Jimmy Chye-Chim Low, who have all done a wonderful job in the production of this book.

As always, I am most grateful to my wife Ruby for typing, retyping, editing, and stylizing, especially the "snacks" strewn throughout the book. She transformed my otherwise sketchy drafts into beautiful stories, the end products of which I barely recognize as my own writing.

CHAPTER 1

Introduction: A Brief History of Entropy and the Second Law of Thermodynamics

In this chapter, I shall present a few historical milestones on entropy and on the Second Law of Thermodynamics. Likewise, I will present a few formulations, interpretations, and popular descriptions of the concept of entropy. The aim of this chapter is not to teach you about entropy (you will have to discover it yourself), nor about the Second Law (you will have to discover this, too), but rather to give you a qualitative description of the kind of problems with which 19th-century scientists were confronted, and from which the Second Law eventually spawned.

There are numerous formulations of the Second Law. We shall classify these into two groups: non-atomistic and atomistic.

1.1. The Non-Atomistic Formulations of the Second Law

Engineers in the late-18th and the 19th centuries were interested in *heat engines*. These engines or machines were supposed to replace human hands or animal power in their tasks. As the term "heat engine" implies, the engine uses heat or thermal energy, to perform useful work.

The simplest way to visualize a heat engine is to start with a waterfall engine. Water cascades from a higher level to a lower level (Fig. 1.1). On its way down, water can rotate a turbine, which in turn can rotate the wheel of a wagon or generate electricity that can be stored for future consumption.

In the 19th century, scientists believed that *heat* was a kind of weightless fluid — called *caloric* — that flowed from a hot body to a cold body. Like the waterfall engine, the caloric fluid "falls" from a higher level temperature to a lower level temperature.

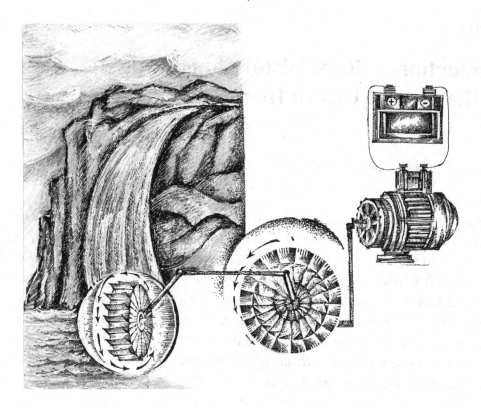

Fig. 1.1   An artist's rendition of a waterfall engine.

On its way down it can either heat a gas in the cylinder or create steam from liquid water (Fig. 1.2). The heated gas or the steam expands its volume. This expansion can push a piston, which in turn can push a shaft, which can lift a weight or rotate the wheel of a wagon, and in the process do useful work for us.

Today, we are familiar with the so-called *internal combustion engine*, where fuel burns *inside* a cylinder, thereby pushing a piston, which then rotates the wheel and runs our cars. The heat engines of the early days can be referred to as *external combustion engines*, which means that the energy supplied to the fluid in the cylinder is provided from the *outside*.

Nowadays, when we talk about the *efficiency* of a car what we actually mean is how far one can get with a given quantity of fuel. Likewise, engineers of the 18th

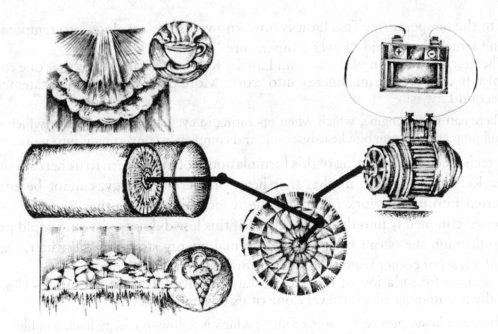

Fig. 1.2 An artist's rendition of a "heat-fall" engine.

century were interested in the *efficiency* of heat engines — how much *work* can be generated from a given amount of heat.

This problem brought to the fore a young engineer named Nicolas Leonard Sadi Carnot (1796–1832). Basing his theory on the concept of the caloric fluid, Carnot discovered that there is an upper limit to the efficiency of a heat engine, and that this limit depends on the two temperatures between which the engine operates.[1] Of course, one can always improve the efficiency of an engine by improving insulation to prevent heat loss, or by reducing friction between moving parts of the engine. However, no matter how "idealized" one makes the heat engine, there is always a limit to the work output one can derive from a given amount of heat that flows from a hot body to a cold body.

Although it was not Carnot who discovered the Second Law, it was his work which sowed the seeds for the inception of the Second Law. Another key discovery that aided the formulation of the Second Law was the *absolute temperature* scale. William Thomson (1824–1907), later known as Lord Kelvin, found that there exists a lower limit to the temperature. This limit is now known as the absolute zero temperature, beyond which there is no "lower temperature."

The first formulation of the Second Law by Kelvin basically states that one cannot completely convert thermal energy into work. More precisely, Kelvin's statement of the Second Law is:

> There can be no engine, which when operating in cycles, the sole effect of which is pumping energy from one heat-reservoir and completely converting it into work.

The precise technical meaning of this formulation is of no concern to us here. In simple terms, Kelvin's statement implies that heat, or thermal energy, cannot be entirely converted into useful work (although the reverse is possible; that is, work can be converted completely into thermal energy). If this law did not exist, one could propel a ship through the ocean by using the thermal energy stored in the water, leaving behind a trail of cooler water. That is, unfortunately, impossible to do.

A second formulation of the Second Law, by Rudolf Clausius (1822–1888), is basically a statement of what every one of us has experienced:

> There can be no process the sole result of which is a flow of energy from a colder to a hotter body.

These two formulations seem to be unrelated on first sight. However, it is easy to show that they are equivalent. If heat could flow spontaneously from a lower temperature to a higher temperature, then one could use the thermal energy stored in the higher temperature to drive a heat engine, and hence violate the Kelvin formulation of the Second Law.

There are of course many other formulations of the Second Law. For instance, a gas confined initially to volume V, if allowed to expand by removing a partition, will always proceed to occupy the entire accessible volume.

One can show that the last formulation is also equivalent to the previous two formulations. There are many other phenomena that we observe in our daily lives — processes that occur in one direction only: Heat always flows from a hot to cold body; a gas will always flow from a smaller volume to a larger one; two gases will always mix in a process as shown in Fig. 1.3.

It was Clausius (Fig. 1.4) who saw the general and unifying principle that is common in all of these processes. His extraordinarily penetrating insight led him to see the unifying principle that "drives" all these processes. Clausius

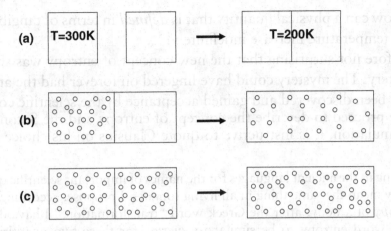

Fig. 1.3 Spontaneous processes: (a) Heat transfer from hot to cold body. (b) Expansion of a gas. (c) Mixing of two gases.

introduced a new term, *entropy*. With this new concept, Clausius was able to proclaim the general overarching formulation of the Second Law of thermodynamics:

> In any spontaneous process, occurring in an isolated system, the entropy never decreases.

Thus, from the large body of experimental observations, a new law was spawned — the law of the ever-increasing entropy, or the Second Law of thermodynamics.

The Second Law brought a new concept into the physicists' vocabulary: entropy.[2] A new concept is always hard to swallow. While most of the terms used in the various formulations of the Second Law, like heat, work, energy, and temperatures, were familiar to physicists, the new term "entropy" was not only unfamiliar but also engendered an element of mystery. Physics was

Fig. 1.4 Rudolf Clausius.

built on conservation laws. Matter cannot be created from nothing and energy is not given to us free. Such conservation laws are perceived as "natural" or as "making sense." But how can a physical quantity that is *defined* in terms of tangible quantities like heat and temperature increase indefinitely?[3]

It is therefore not surprising that the new concept of entropy was shrouded in a cloud of mystery. The mystery could have lingered on forever had the atomic theory of matter not been discovered and gained acceptance by the scientific community.

Before we proceed to describe the concept of entropy and the Second Law in its atomistic formulation, it is instructive to quote Clausius on his choice of the word "entropy":

> I prefer going to the ancient languages for the names of important scientific quantities, so that they mean the same thing in all living tongues. I propose, accordingly, to call S the *entropy* of a body, after the Greek word "transformation." I have designedly coined the word entropy to be similar to *energy*, for these two quantities are so analogous in their physical significance, that an analogy of denominations seems to me helpful.[4]

It is not less instructive to quote Leon Cooper's comment on Clausius' choice of the word "entropy":

> By doing this, rather than extracting a name from the body of the current language (say: *lost heat*), he succeeded in coining a word that meant the same thing to every-body: *nothing*.[5]

It is true that the word "entropy" meant "nothing" not only to a lay person but also to many scientists who were working in the field of thermodynamics. One can *define* entropy, and *measure* entropy changes using familiar quantities such as heat and temperature, but its meaning remains a mystery and eludes interpretation within the realm of macroscopic thermodynamics. I shall add here two personal comments regarding the *term* entropy. First, Clausius erred in his statement that entropy is *similar* to energy and that these two quantities are *analogous* in their physical significance. Second, although I agree with Cooper that entropy means *nothing*, I do not share with him the view that "lost heat" could be a better term than entropy. For more details see Ben-Naim (2008).

1.2. The Atomistic Formulation of the Second Law

Towards the end of the 19th century, three branches of physics — mechanics, electromagnetism, and thermodynamics — reached a high degree of theoretical perfection.

These three branches seem to have dealt with three completely unrelated physical entities. All three theories were applied to *macroscopic measurable* quantities. Mechanics dealt with the relationship between the motions of objects and the forces exerted on them. Electromagnetism combined electrical and magnetic phenomena with electromagnetic radiation, such as light. Thermodynamics focused on heat engines and the relationship between heat (or the caloric fluid), energy and useful work.

Each of these disciplines was founded on a small number of laws and was proclaimed to be all-encompassing and absolute. These laws were formulated without explicit reference to atoms and molecules. Furthermore, the mathematical formulations of these laws were devoid of any tinge of probability or randomness.

The first breakthrough in the emergence and eventual acceptance of the atomic theory of matter was the development of the kinetic theory of gases. The theory succeeded in explaining the properties of gases based on the assumption that the gas consists of small particles: *atoms* and *molecules*. Atoms and molecules were only hypothetical entities envisioned by Greek philosophers more than two millennia ago. The macroscopic quantities that we measure in our laboratories, such as heat and temperature, were now *interpreted* in terms of the motions of the atoms and the molecules. However, *interpretation*, no matter how sound it may be, does not provide a recipe for *calculating* measurable quantities. All the existing theories of physics could be successfully applied to a small number of objects. On the other hand, accepting the atomistic constituency of matter entails dealing with a huge number of particles. To

Fig. 1.5 James Clerk Maxwell.

relate the macroscopic measurable quantities to the atomic motions of particles, one was compelled to use *statistical methods*, and statistical methods require the tools of *probability*. The main protagonist on the scene who introduced statistical methods into physics was James Clerk Maxwell (1831–1879) (Fig. 1.5).

In spite of the remarkable success of the kinetic theory of gases to explain the properties of gases, the very existence of atoms and molecules was still considered to be a hypothesis only: No one had ever *seen* an atom or a molecule. Physicists and chemists were divided in their attitude towards the atoms. Some who accepted the atoms were cautious and hesitant, while others adamantly refused to accept the atoms as real physical entities.

It should be mentioned that in the kinetic theory of gases, the probabilities were used only as auxiliary tools to calculate *average* quantities such as volume, pressure, and temperature. Accepting an *average* quantity as a meaningful physical quantity is one thing, and accepting the usage of *probability* theory in physics is quite another. There was a profound uneasiness in physics with the notion of probability itself as a major player in physical phenomena.

Fig. 1.6 Ludwig Boltzmann.

From this background a giant physicist, Ludwig Boltzmann (1844–1906), stepped forward (Fig. 1.6). Boltzmann was a staunch supporter of the atomic theory of matter and, together with Maxwell, contributed to perfecting the kinetic theory of gases. Yet, his boldest and most far-reaching contribution to physics was his atomistic formulation of entropy and the Second Law of thermodynamics.

Boltzmann's formulation of entropy seems to come from nowhere. It relates the entropy — that quantity which Clausius had *defined* in terms of heat and temperature — to the *number of states of the system* (Fig. 1.7). For the purpose of this introduction it does not matter whether you know what the "number of states" means, just as it does not matter what heat or temperature means. It is clear that there is a huge conceptual abyss between Clausius' definition of entropy in terms of two *physically measurable quantities* of heat and temperature on one hand, and the *number of states*, or configurations, or arrangements of a system consisting of a large number of atoms, on the other hand.[6] The

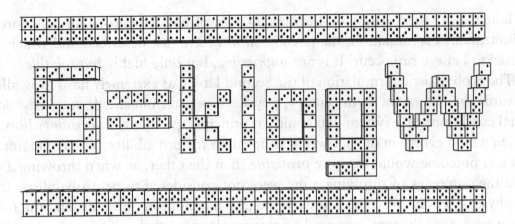

Fig. 1.7 Boltzmann's entropy (rendition with dice).

latter seems to be devoid of any physics at all. Nevertheless, Boltzmann's formula (Fig. 1.7) became one of the cornerstones of modern physics.

In Boltzmann's formulation of the entropy, the main protagonist was the *probability* itself, not *average* quantities that can be computed by the tools of probability theory.

The idea that probability is the "driving force" behind the Second Law threatened to undermine the very foundation of the Second Law, which was proclaimed as an absolute law of nature without provisions for any exceptions.

If you look at a glass of water into which you have just added a few ice cubes, you will see shortly that the ice cubes will melt and the water will get colder. This seems to be a *natural* phenomenon. It will always occur in this particular order of events. This is one manifestation of the Second Law.

What if I told you that I was looking at a glass of water and I saw suddenly how ice formed in one part of the glass while in the other part the water became very hot. Would you believe me? Certainly not. Why?

Because you know that this is against the Second Law. Even if you have never heard of the Second Law, you will not buy my story because no one has ever observed such a phenomenon. Therefore, the story I told you sounds *unnatural* or an impossible event.

Boltzmann's formulation of the Second Law was based on probability. According to Boltzmann's formulation, the phenomenon I have just described and which you refused to believe *can* occur. It is not *impossible*, but only highly *improbable*.

The probabilistic formulation of the Second Law was extremely hard to swallow. Towards the end of the 19th century, physics was deterministic. If you knew some initial conditions of a system you could, in principle, predict with *certainty* how the system would evolve in time. There was no room for probability. You could not say that one outcome would be more *probable* than the other, as when throwing a die. Even the outcomes of throwing a die were not considered to be probabilistic from the physicist's point of view. Knowing the initial condition of the die and the forces acting on it at each stage, one could, in principle, predict the outcome at any given time. The involvement of probabilities in games of chance was explained as a result of our lack of a complete knowledge of the initial state, and of all the details of the forces acting on the die. Probability was not part of physics, certainly not the kind of stuff on which to found a law of physics. A law of physics must be absolute: No exceptions are allowed!

Not surprisingly, Boltzmann's formulation was not accepted easily, not only by those who rejected from the outset the atomistic theory of matter but also by those who believed in the atoms but could not reconcile with the invasion of probability into physics.

All of this had occurred before the advent of quantum mechanics, in which probability became the central pillar of physics and probabilistic thinking pervaded the minds of all physicists.

In the following chapters you will discover how the Second Law emerges from probabilistic behavior of a system of many particles. However, before doing that we have to complete our story of entropy by introducing the third, and a conceptually very different, view of entropy.

1.3. Entropy as a Measure of Information

The title of this section is not a joke! Even if you have never heard of entropy before, and have read only the two previous sections, you would not be able to understand how entropy is associated with information. Accepting entropy as originally defined in terms of *measurable* quantities like heat and temperature,[3] you would be bewildered to read that the same concept, entropy, is also associated with such a nebulous and intangible concept like information.

As I promised you in the preface of this book, you already know what probability is. You certainly know what information is. Moreover, I shall show you in Chapter 2 that you also know what a "measure of information" is. But before doing that we must continue with the story of entropy.

Information is a very general concept. The information contained in the previous sentence is not new to you. You might not know how to *define* information; neither do I. But you certainly know what it is, and you also know that it is "a very general concept."

"Tomorrow, it will rain at noon."
"This book was written in Sweden."
"I believe you are enjoying reading this book."

These sentences convey *information*. We shall expound more about that in the next chapter. Here, we note that information in itself is *irrelevant* to entropy and the Second Law.

While working at Bell Laboratories, Claude Shannon (1916–2001) was engaged in developing a theory of communication. The problem he tackled was how to transmit *information* through communicating channels in the most efficient way. In 1948, he published his seminal work, *A Mathematical Theory of Communication*.[7] In this work,

Fig. 1.8   Claude Shannon.

Shannon describes in a clear and detailed way what kind of a *measure* of *information* he was seeking and what he eventually found (Fig. 1.8).

As we shall see in the next chapter, not all types of information can be measured. But some can be, and the measure Shannon found became the cornerstone of information theory.[7] Nowadays, Shannon's measure of information (*SMI*) is used in such diverse disciplines ranging from statistical mechanics and communication theory, to economics, linguistics, psychology, music, and a host of others.

Shannon's measure of information (*SMI*) is a very general and abstract concept. In my view, it is also a beautiful concept. Its extraordinary interpretive power has helped to fathom the meaning of entropy. We shall encounter *SMI* in connection with the 20-questions games in the next chapter, and we shall use *SMI* in the rest of the book to follow some simple experiments performed with marbles in boxes or cells.

It is one of the unfortunate events of history that *SMI* got a misleading and an incongruous twist. The story as told by Tribus (1971) is as follows:

> What's in a name? In the case of Shannon's measure the naming was not acciden-tal. In 1961 one of us (Tribus) asked Shannon what he had thought about when he had finally confirmed his famous measure. Shannon replied: "My greatest con-cern was what to call it. I thought of calling it 'information,' but the word was overly used, so I decided to call it 'uncertainty.' When I discussed it with John von Neumann, he had a better idea. Von Neumann told me, 'You should call it entropy, for two reasons. In the first place your uncertainty function has been used in statistical mechanics under that name. In the second place, and more impor-tant, no one *knows what entropy really is, so in a debate you will always have the advantage.*' "[8]

As I have said above, the adoption of the term "entropy" for *SMI* was an unfortunate event, *not because entropy is not SMI, but because SMI is not entropy*!.

You will recall that Clausius coined the term "entropy" within thermodynamics.[3] In my view, that term was an unfortunate one for the quantity that Clausius had discovered. It is *a fortiori* an unfortunate term to be used in all those fields where the *SMI* is successfully applied but somehow perversely referred to as entropy.

The association of entropy with some kind of information has its roots in Maxwell's writing, but that was only implicit. In 1930, GN Lewis wrote:

> "Gain in entropy always means loss of information, and nothing more."[9]

That statement uses the word *information* explicitly, but in its qualitative colloquial sense. Now that we have Shannon's measure of information, we no longer need to identify entropy with information *itself*, but as a particular case of *SMI*.

If all that I have written in this section seems quite abstract, do not fret. In the next chapter we shall discuss both the concepts of probability and *SMI* in terms of simple games. We shall continue to use these concepts throughout Chapters 3–6. In

Chapter 7, we shall translate our findings into the language of real systems consisting of a huge number of particles.

1.4. Popular Metaphoric Descriptions of Entropy

If you open any dictionary and look at the meaning of entropy, you might find terms like "measure of disorder," "mixed-upness," "unavailable energy," and many others.

Here is a quotation from the online free dictionary.

Entropy

1. *Symbol S.* For a closed thermodynamic system, a quantitative measure of the amount of thermal energy not available to do work.
2. A measure of the disorder or randomness in a closed system.
3. A measure of the loss of information in a transmitted message.
4. The tendency for all matter and energy in the universe to evolve towards a state of inert uniformity.
5. Inevitable and steady deterioration of a system or society.

Unfortunately, none of the above is a correct definition of entropy. It should be noted that Boltzmann himself was perhaps the first to use the "disorder" metaphor in his writing:

> ... are initially in a very ordered — therefore very improbable — state ... when left to itself it rapidly proceeds to the disordered most probable state.[10]

You should note that Boltzmann uses the terms "order" and "disorder" as qualitative descriptions of what goes on in the system. When he *defines* entropy, however, he uses either the *number* of *states* or probability.

Indeed, there are many examples where the term "disorder" can be applied to describe entropy. For instance, mixing of two gases is well described as a process leading to a higher degree of disorder. However, there are many examples for which the disorder metaphor fails.[11]

Look at the two illustrations on the first pages of this book and enjoy the artful drawings of Alex Vaisman. If you have been exposed to the association of entropy with disorder, you might be puzzled by the titles "Low entropy room" and "High entropy room." You might also suspect that there is an error in these titles. I hope that by the time you finish reading this book, you will understand why I have chosen these titles.

Recently, usage of "spreading of energy" as a metaphor for entropy seems to be in vogue. As we shall see throughout the book, both disorder and the spreading metaphors have some merits in describing what goes on in the system. However, none can be used to define or even to describe precisely what entropy is. It seems to me that there exists some confusion in the literature, between a *description* of what happens in a spontaneous process on the one hand, and the *meaning* of entropy on the other hand. We shall further discuss these terms after we have gained some experience with simple "spontaneous" processes in games of marbles in cells. You will be able to decide for yourself when one description is valid and when it is not. You will also see that *SMI* is not a metaphor for entropy — *entropy is a particular example of SMI.*

Snack: Who's Your Daddy?

One fine spring day, Claude went strolling near the Princeton University Campus. He was not alone enjoying that lovely spring day, however. His newborn baby, who was in a baby carriage, was with him. He started to play with the baby and was both deeply engrossed and amused at the baby's progress when he felt a light tap on his shoulder.

"Hello there! How does it feel to be a brand-new Daddy?" his friend asked. "You startled me! Hi, John! It's good to see you, and to answer your question, it's great to be a Daddy. My son is such pure joy," Claude proudly proclaimed.

"What's his name?" John asked.

"We have not decided yet. We have two choices, Information or Uncertainty. My wife and I think that either one of these is appropriate for our son, but my wife thinks these are overused names."

John took a peek at the baby carriage and he was indeed convinced that the baby could very well be anyone's pride and joy. However, there was something that struck him. The baby had a striking resemblance to someone he had seen a long time ago.

Fig. 1.9   Claude and John in the park.

After a few seconds of googling into his memory, the thought struck him: "Why yes, of course, this baby looks so much like Rudolph's son, Entropy."

"Hey Claude, why don't you call your son Entropy?" John suggested. "The names you are considering are indeed over-used, but Entropy is a rare name and perhaps even a sexy one, and no one in the campus is named Entropy and no one understands what it means either. I think Entropy would be a nice name for your baby boy, and besides he looks like Rudolph's son," John pressed on.

Claude was intrigued by this name and, after discussing it with his wife, they agreed to call the baby Entropy.

As the young Entropy grew up, everyone who heard his name was fascinated by it. The name Entropy became so popular that scientists and non-scientists, economists, mathematicians, psychologists, and artists the world-over used it when naming their own babies.

Many were confused between the two sons named Entropy; some misused and some even abused the name. Some people who also knew Rudolph's son suspected that Entropy, Claude's son, was actually Rudolph's and not Claude's. Rumors spread like wildfire and, in no time at all, many people who met Entropy, Claude's son, thought that he was Rudolph's son.

Claude was irritated by the continual misuse of his son's name. One day he met his friend Heira and told him the story of his son's name. Upon hearing the story, Heira suggested to Claude to change his son's name to a more meaningful name, one that did not conjure all kinds of mysterious gossip, and also a name such that no one would confuse his son with Rudolph's son. Heira then said: "Why don't you call him Smomi, a short acronym for 'Shannon's measure of missing information.' That cute name can be further abbreviated to *SMI*. With this name, it will be clear once and for all who the real biological father of your son is."

Claude liked it, but was hesitant to change his son's name.

The irony of it all was that years later, after an intense study of the DNA of Rudolph's Entropy, they found convincing evidence that Entropy — Rudolph's son — was in fact Claude's son. The striking resemblance between the two sons which was first noticed by John was due to this fact, and not the other way around.

CHAPTER 2

All You Need to Know but Never Dared to Admit that You Already Know

In this chapter, we shall discuss the prerequisites for understanding the rest of the book. The two fundamental concepts — probability and information — are discussed in Sections 2.2 and 2.3, respectively. Upon each of these concepts, well developed and highly sophisticated branches of mathematics have been erected.

Hold on! Do not cringe upon hearing the words "highly sophisticated mathematics." You can relax. None of this mathematics will be needed for understanding this book. As stated in the chapter title, you already know all that there is to know. The purpose of this chapter is meant to convince you of the veracity of its title. Of course, if you aspire to go beyond the mere understanding of what entropy is and why it changes in one direction only, you will have to learn much more. But that is beyond the scope of this book.

How do I know that you are already acquainted with the fundamental concepts of probability and information? The answer lies within the pages of this chapter.

Sections 2.2 and 2.3 both begin with a short aperitif consisting of some simple games that you are familiar with, and will hopefully facilitate your digestion of the main course that follows. The main course consists of a more in-depth analysis of the games we played in the aperitif; what was the game's purpose, what have we learned from the game, and what conclusion can be drawn from these games?

Finally, the dessert. Like any real dessert, it is meant to be pleasurable and savored. It will also grant you further reassurance that you have the ability to understand the rest of the book.

In Section 2.1, we shall discuss several simple and helpful concepts. First, grouping of objects and assigning a measure; second, the logarithm of a number, especially of very large numbers; and third, the distinction between a specific event (or a configuration or an arrangement), and a dim-event. These concepts are very simple and easy to grasp. You will also have to know that matter consists of atoms and molecules. That is just about all you need to know about physics.

2.1. Grouping of Objects and Assigning Measures

Aperitif: Playing Lottery

You are given a choice between two lotteries, (a) and (b) (Fig. 2.1).

(a) There are six prizes. The value of each is 10 US$ dollars but all the prizes are in different currencies, shown as different colors in Fig. 2.1a. There are also six prizes of 1 US$ each in US currency (shown in gray).

(b) The second lottery offers 12 prizes of 10 US$ dollars but all are in different currencies. There are also six prizes of 1 US$ in US currency (shown in gray).

You are given one free ticket to participate in either one of the two lotteries. Answer the following two questions:

1. Which lottery will you choose presuming that you are *specifically* interested in winning the bill in Indian currency?

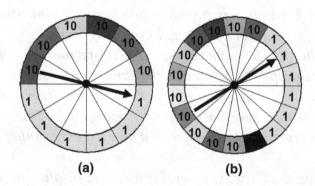

Fig. 2.1 Two lotteries as described in Section 2.1. The colors represent different currencies. The numbers are the values of the prizes.

2. Which lottery will you choose if you are interested in maximizing your gain, regardless of currency?

Think about the two questions, answer them first then compare your answers with my answers, provided in Note 1.

The Main Course: Hierarchy of Objects, Grouping and Assigning Weights, Very Large and Very Small Numbers

This course consists of several topics that will be helpful in understanding the rest of the book.

First, let us "digest" the aperitif. This aperitif was designed to demonstrate the process of grouping objects according to some criterion, or some rule, or some property. In this specific example we made a distinction between a *specific* prize — say, the Indian currency or the Chinese currency — and the group of prizes that we deem to be equivalent; here, all the prizes, the values of which are 10 US\$. We can refer to the first case as a *specific* description (of the prize), and the second case as a *dim* description (of the prize). Clearly, our choice of lottery depends on whether we are interested in a *specific* prize or a *dim* prize.

Let us move to the next example of grouping objects, but now we shall also assign a *measure* to each of the objects.

Consider all possible physical objects: a person, an ocean, a cloud, a specific cat, a book, the specific book you are holding now, and so on.

Clearly, the set of all of these objects is extremely large. Here, we are not interested in the objects themselves but in some measure *defined on* the object — examples are the volume or the weight of the objects. Clearly, not all of the objects you can imagine can be assigned a volume or a weight — a cloud has no definite boundaries and therefore it is difficult to assign a volume or weight to it. The ocean is ever-changing, it has no clear-cut boundaries as water ebbs and flows, and it is not clear which water to include in weighing the ocean. A book is an abstract object, but a specific book (e.g. the one you are holding right now) can be assigned a weight.

Let us build a hierarchy of groups or sets of objects. Before doing so it is important to know that all matter consists of molecules, and molecules consists of atoms of different elements. There are about 100 elements; for instance, diamond is pure carbon, liquid mercury consists of mercury atoms, and neon gas consists of neon atoms. The fact that all matter consists of atoms and molecules is one of the most important — if not *the* most important — discovery of science.

Here, we are interested only in the existence of atoms and molecules, and that to each atom a measure such as its mass or its weight may be assigned.

The first group consists of all objects which are made of pure elements; that is, they consist of atoms of the same kind. Examples are a block of pure gold, a diamond, or liquid mercury. Since each atom can be assigned a weight, an object that is a pure element will be assigned the total weights of all its atoms. This is simply the weight that you will be measuring with a balance. We call this group of objects *pure elements*.

A molecule consists of a combination of atoms. The molecule can also be a pure element; for instance, oxygen molecules in the air consist of two atoms of oxygen. But a water molecule consists of one oxygen and two hydrogen atoms. Since each atom is assigned a weight, a molecule can also be assigned a weight, simply the sum of the weights of its constituent atoms.

The second group of objects consists of all objects that are *pure substances*; that is, they consist of one type of molecules. For example, a block of ice consists of water molecules; a drop of ethanol consists of molecules of ethanol, each of which consists of one oxygen, two carbon, and six hydrogen atoms. Since each molecule is assigned a weight, each object of pure substance can also be assigned a weight — the sum of the weights of all the molecules in that object.

Obviously the group of all objects that are pure substances is larger than the group of all objects that are pure elements. Not only larger, but the first *includes* the second group. This is shown schematically in Fig. 2.2.

As shown in Fig. 2.2, the set of all points representing objects that are pure elements is a subset of all objects that are pure substances. Any object which is a pure element is also a pure substance. The converse of this statement is not true. Every object that is a pure substance is not a pure element.

Finally, let us take all possible physical objects — the book you are holding, the pen on your desk, and so on. Clearly, this group of objects is immense. It includes the subset of all objects that are pure substances, which includes all objects that are pure elements.

What we have achieved so far is that we started with the general concept of physical objects, each of which may or may not be assigned a measure such as weight

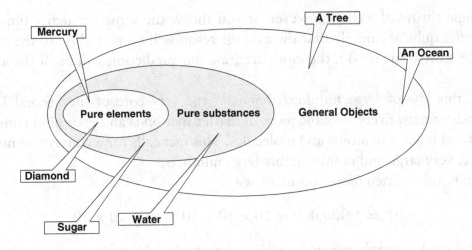

Fig. 2.2   Grouping of objects into inclusive sets.

or volume. We then selected a subset of this immense set of physical objects; all physical objects that are pure substances. Each of these objects can be assigned a weight. From this subset we have selected yet another subset; all objects that are pure chemical elements. Each of these specific objects also has a definite weight.

We shall repeat this process of grouping into subsets in Sections 2.2 and 2.3. Instead of physical objects, however, we shall group *events* in Section 2.2, and types of *information* in Section 2.3.

Before concluding this part of our course it should be noted that there are many ways of grouping objects into different subsets (e.g. all objects that are tables, a subset of all objects that are furniture, a subset of all physical objects). There is in general more than one measure that can be assigned to objects. Here we chose the weight for illustration, but we could have chosen volume, surface-area, etc.

The Second Course: Extremely Small and Extremely Large Numbers

We now move on to discuss numbers; very large and very small.

One of the most amazing results of the theory of probability is that if you repeat an experiment (or a game) a multitude of times, some regularities emerge in spite of the fact that each outcome of a single experiment (or game) is quite random and unpredictable. A simple example is throwing a die. In a single throw, the outcome can be any number between 1 and 6. There is no way of predicting the outcome of a single throw of a die. However, if you throw the same die many times, you can *predict* quite accurately that the average result will be very close to the number 3.5! The more trials we do, the more accurate our prediction will be of the average result.

It is this law of *large numbers* that is at the very core of the Second Law of Thermodynamics. Every tangible piece of matter that one can see or hold consists of billions and billions of atoms and molecules. This fact calls for a convenient notation for large, very large and unimaginably large numbers.

A million is written more succinctly as:

$$10^6 = 1,000,000 = 10 \times 10 \times 10 \times 10 \times 10 \times 10$$

The notation 10^6 simply means take 10 and multiply it by itself 6 times — the result is one million. Clearly, one million can be written explicitly as 1,000,000. But what about a million, million, million, million, millions? This entails writing a whole line of zeros explicitly. It is therefore more convenient to write this number as 10^{30}, which is simply a shorthand notation for 1 followed by 30 zeros, or the result of multiplying 10 by itself 30 times.

The number 10^{30} can still be written explicitly, as 1 followed by 30 zeros. This is quite a long string of zeros, but can still be written.

In thermodynamics we encounter numbers of the order $10^{(10^{10})}$. Simply put, this means multiplying 10 by itself 10^{10} times, or 1 followed by 10^{10} zeros. Can you imagine how large this number is? Try to write this number explicitly:

$$1,000,000,000,000\ldots$$

Suppose you can write 6 zeros in a second. You will write 6×60 zeros in a minute; in an hour you will write $6 \times 60 \times 60$ zeros; in 24 hours you will write $6 \times 60 \times 60 \times 24$ zeros; in a year you will write 1 followed by:

$$6 \times 60 \times 60 \times 24 \times 365 = 18,921,600 \text{ zeros.}$$

Thus, even if you go on writing for a whole year, you will still be far away from your goal of writing 10^{10} zeros. Even if you continue writing for another 50 years, writing a seemingly endless string of zeros, you might finish that task and obtain a number having $6 \times 60 \times 60 \times 24 \times 365 \times 50 = 9,460,800,000$ zeros. This is "close" to the goal number.

In thermodynamics, we encounter numbers that are even larger than those we tried to write explicitly, numbers of the order of $10^{(10^{30})}$. Such numbers are beyond our imagination. They cannot be written even in billions and billions of *years* — far longer than the estimated age of the universe.

Thus, in order to calculate the probabilities and numbers of configurations (that is what we shall be doing in Chapters 3–5), we need a more convenient notation. Instead of explicitly writing the number 10^{10} we shall simply write the number 10 and call it the *logarithm* of the number 10^{10} to the *base* 10, and denote it by $\log_{10}\left(10^{10}\right)$. This is simply the number of times you have to multiply 10 by itself to get the number 10^{10}. Similarly, $\log_{10}\left(10^{10^{10}}\right)$ is the number 10^{10}. You have to multiply the *base* 10, 10^{10} times. This is simply a shorthand notation for large numbers, nothing more.

Instead of the *base* 10, it is sometimes convenient to use the *base* 2. The logarithm to the base 2 of the number 2^{10} is simply $\log_2 2^{10} = 10$. This is the number of times you have to multiply the number 2 by itself to get the number 1024; that is:

$$2^{10} = 2 \times 2 \times 2 \times 2 \times 2 \times 2 \times 2 \times 2 \times 2 \times 2 = 1024$$

This is a large number, but still a manageable, number but $2^{(2^{30})}$ cannot be written explicitly. Instead, we write the logarithm to the base 2 of this number, which is:

$$\log_2\left(2^{(2^{30})}\right) = 2^{30} = 1,073,741,824$$

The concept of logarithm can be applied to any number, not only to numbers of the form 10^{10} or 2^{10}, but we shall not need this for the present book.

There is one very important property of the logarithm which is useful to know. The logarithm of a *product* of two numbers is the *sum* of the logarithm of the two numbers. You can easily verify that with an example:

$$10^5 \times 10^3 = (100,000) \times (1000) = 100,000,000 = 10^8$$

Therefore, the logarithm of the product $10^5 \times 10^3$ is the sum of the logarithm of 10^5 and the logarithm of 10^3, which is 8.

The Dessert: Young Children Know How to Group Objects Playing an "Unfair" Roulette

This dessert consists of two parts. The first consists of some experiments performed with children (aged 6–16) that demonstrate their ability to group different things and treat them as "the same" or "alike." The second is an imaginary experiment with children of any age, playing an "unfair" die.

a. Grouping of objects

Olver and Hornsby[2] reported on the results of extensive research on children's ability to group different objects and treat them as being equivalent in some sense. They also studied the basis on which a child renders things as being equivalent. Such equivalence-making is in large measure a learned capability and, as expected, it may change with the child's growth and development.

Fig. 2.3 The original pictures used in the equivalent task with pictorial material from the article of Olver and Hornsby (1966).

I will present here only one experiment of "Equivalence formation with pictures," reported by Olver and Hornsby (1966).[2] We shall require similar "pictures" in Section 2.3.

Children aged 6–11 were shown an array of 42 watercolor pictures (Fig. 2.3). The children's task was to select from this array a group of pictures that were "alike in some way." All the drawings were of familiar objects, such as a dog, a pair of scissors, a doll, etc. The children were first asked to identify the objects in the pictures to ensure that they were familiar with them. The next step was to have them choose the pictures that were "alike in some way — any way at all in which a group of things is the same."

After having completed the grouping, the children were then asked *how*, or in what sense, the selected objects were alike.

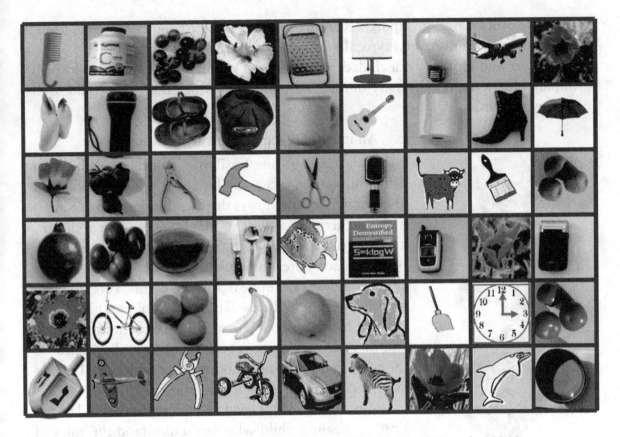

Fig. 2.4 Array of objects similar to that used by Olver and Hornsby (1966).

There are, of course, many ways of grouping objects in sets of equivalence according to different attributes. For instance, objects that are tools, or are of the same color, or are food (Fig. 2.4).

In these particular experiments, Olver and Hornsby studied the relative use of *perceptible* and *functional* attributes as the basis of the grouping task. They found that across all ages, children used perceptible properties as the basis for judging "likeness." The younger children were more likely to use the perceptible properties (such as colors, size, and shape) for grouping than the older ones. The tendency to use perceptible attributes declines systematically with age.

On the other hand, grouping according to functional attributes (such as tools, means of transportation, etc.) typically increases with the child's age.

As Olver and Hornsby concluded, were children not naturally prone to do such grouping of objects, the diversity of the environment in which they live would soon overwhelm them.

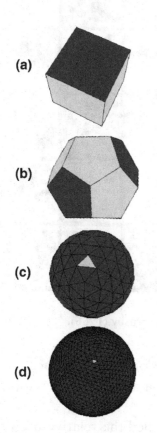

(a)

(b)

(c)

(d)

Fig. 2.5   The unfair dice.

b.   An imaginary experiment with an unfair die

I show a child of age Epsilon a multifaceted die with different-colored faces, say blue and yellow (Fig. 2.5). I also show him how I toss the die, doing so by tossing it high and giving it a twist so that it spins while in the air, and that when the die lands it may reveal either a yellow or a blue face. After showing him how it works, I ask the child to choose either one of the two colors. I also explain to him that should the die land on the color of his choice, he shall have earned a bar of chocolate whereas, on the contrary, he would not get anything if the die landed on the color that he did not choose. After each game I would ask the child why he made that choice.

Before I continue, I need to explain that a child of age Epsilon means a child who can understand the rules of the game as outlined above. The reader is encouraged to conduct this experiment or at least to imagine conducting this experiment with a child of any age.

In the first experiment I used a regular die with six faces, as in Fig. 2.5a. The total blue and yellow surface areas were equal. The child chose yellow. I threw the die and it showed a blue face. The child was disappointed. I asked him why he chose yellow. He hesitated to answer and said that he did not get the question, and that he did not know why he chose that color.

The experiment was repeated with the same die (a). This time the boy chose blue. I asked why he chose blue. His immediate reply was that yellow was not good and that it failed to deliver the prize. I tossed the die again and it landed on a blue. This made the child very happy.

The next die was (b) (Fig. 2.5b). The number of yellow faces was twice as great as the number of blue faces. The child chose yellow. When asked why he chose the same color that failed him in the last game, he responded by saying that there are more yellow faces. In the third game, I chose a die with the blue color about 20 times greater than the yellow-colored faces (Fig. 2.5c). This time the child switched to the blue color. Once again, I asked the child why he chose that color. Without the benefit of an answer, the child's expression gave away his thoughts. Prodding the child further for an answer, he said that with yellow he will *never* win.

Clearly, the child has a sense of what is bigger and, when presented with a choice, he or she will not hesitate to choose the dominant color.

The last game was with a die that I did not have. I showed a picture of it to the child (Fig. 2.5d) and explained that this die has billions and billions and billions of faces. One of the faces is yellow and all the rest are blue. I repeated the game. The child obviously chose blue and, when asked as to why he chose that color, with an impish grin he said: "Are you kidding?"

In the next section, I shall describe a few, slightly more sophisticated experiments to probe children's perception of probability. In this section, I wanted only to accentuate the sense of the relative sizes of two events. When the relative size is very large, a child does not hesitate to use the words "never" or "always" although, strictly speaking, these terms are not correct. We shall see that the same words can be safely applied in connection with the Second Law.

2.2. Events and Their Probabilities

The Aperitif: Playing with Dice

To get into the groove, let us start with a warm-up game. You choose a number between 1 and 6 — say 4. We throw a fair die (Fig. 2.6). If the outcome is 4, you get 4 US\$, but if the outcome is not 4, you have to pay 1 US\$. Which number will you choose, assuming that we will play the game 1000 times and that you want to maximize your earnings?

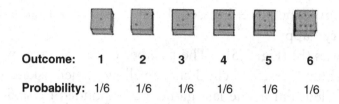

Outcome: 1   2   3   4   5   6

Probability: 1/6   1/6   1/6   1/6   1/6   1/6

Fig. 2.6 A fair die and the probabilities of the outcomes.

I am confident that even without giving it a second thought, you know that it does not matter which number you choose. On average you will always lose in this game, whichever number you choose. Let us calculate your chances. Whichever number you choose, the probability of its occurrence is 1/6. If you get 4 US$ each time your chosen number has occurred, and you played 1000 games, your expected earnings are:

$$\frac{1}{6} \times 1000 \times 4 = \frac{4000}{6} = 666.66 \text{ US\$}$$

On the other hand, the probability of the occurrence of a number that you did not choose (i.e. 1, 2, 3, 5, 6) is 5/6. If you pay 1 US$ on each "unlucky" outcome, then your expected loss in 1000 games is:

$$\frac{5}{6} \times 1000 \times 1 = \frac{5000}{6} = 833.33 \text{ US\$}$$

Thus, no matter what you have chosen, after playing this game 1000 times you will lose about $833 - 666 = 167$ US$.

Let us proceed to a game with two dice. You choose a number between 2 and 12. These are all the 11 possible *sums* of the outcomes of two dice. If the sum you have chosen comes up, you get 8 US$. If the sum is different from what you have chosen then you pay 1 US$. We agree to play the same game 1000 times. You are allowed to choose a number between 2 and 12 only once during the entire game. Assuming that you want to maximize your earnings, what number should you choose?

This game requires some thinking. The first thing you should realize is that unlike the previous game, here each choice of a *sum* has a *different* probability of occurrence. Therefore, it matters what number you choose!

Let us do some simple calculations. It is very important that you follow the calculations in this relatively simple example. Later, we shall encounter similar problems with a similar *type* of calculation, but with much bigger numbers.

In Fig. 2.7 we show all possible *specific* results of throwing two dice. A *specific* result means that we *specify* which die shows which number. Altogether, we have 36 possible outcomes. Check Fig. 2.7 to make sure we did not miss any possible result and that we did not count any specific result more than once.

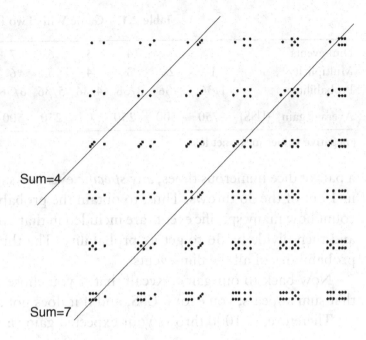

Fig. 2.7   Sum of the outcomes of two dice.

When we choose the *sum* of the outcome we do not care about the specific number on a specific die. All we care about is that the *sum* of the outcomes will be a number between 2 and 12. For instance, the *event*: "the sum of the outcomes is 4" does not specify which number is on which die. In order to obtain this event we need to collect all the outcomes in Fig. 2.7 into groups. In each group the sum of the outcomes is the same. Clearly, the event *sum* = 4 comprises three specific events:

$$1:3, \quad 2:2, \quad 3:1$$

We shall refer to the event *sum* = 4 as a dim-event.

We shall call the number of *specific* events that comprise a dim-event the *multiplicity* of the dim-event. In Table 2.1 we list all possible dim-events, their multiplicities, and their probabilities.

What is the probability of the event *sum* = 4? In total, we have 36 specific events. Therefore, the probability of each *specific* event is 1/36. This means that if we throw

Table 2.1.   Game With Two Dice

Dim-events	2	3	4	5	6	7	8	9	10	11	12
Multiplicity	1	2	3	4	5	6	5	4	3	2	1
Probability	1/36	2/36	3/36	4/36	5/36	6/36	5/36	4/36	3/36	2/36	1/36
Average gain* (US$)	−750	−500	−250	0	250	500	250	0	−250	−500	−750

*negative values mean net loss.

a pair of dice numerous times, any *specific* event — say 3:1 — will appear on average in 1 out of the 36 throws. Thus, to obtain the probability of a dim-event we have to count how many specific events are included in that event — that is, its multiplicity — and then divide by 36 to get its probability. The third row in Table 2.1 shows the probabilities of all the dim-events.

Now back to our game. Recall that if you chose a sum — say, *sum* = 4 — and that sum appears, you earn 8 US$, and if it does not appear you lose 1 US$.

Therefore, in 1000 throws your expected gain is:

$$\textit{probability} \times 8 \times 1000 = \frac{3}{36} \times 8 \times 1000 = 666.66 \text{ US\$}$$

On the other hand, the probability of getting a result other than *sum* = 4 is 33/36 (i.e. the number of *specific* results in Fig. 2.7 for which the sum differs from 4 is 33). Therefore, in 1000 throws your expected loss is:

$$\frac{33}{36} \times 1 \times 1000 = 916.66 \text{ US\$}$$

Thus, if you play the same game 1000 times with the choice of *sum* = 4, you will lose about 916 − 666 = 250 US$.

Can you fare any better? Obviously you are not going to choose *sum* = 2 or *sum* = 12. Why? Because these sums have very low probability of occurrence and you stand to lose much more than in the case of choice *sum* = 4. Looking at the third row in Table 2.1, you will see that *sum* = 7 has the highest probability, namely 6/36. Let us calculate your expected gains if you choose *sum* = 7:

$$\frac{6}{36} \times 8 \times 1000 = 1333.33 \text{ US\$}$$

And the expected loss:

$$\frac{30}{36} \times 1 \times 1000 = 833.33 \text{ US\$}$$

You see that in this case your average gain is about 1333 − 833 = 500 US\$. Not bad at all for a small investment which requires only a little thinking!

Exercise E2.1: Calculate the average gain or loss for the choice of *sum* = 3 and *sum* = 11. Explain why the results are equal.

It is easy to calculate the average expected gain (a minus sign means a loss) for each choice of a sum. If you choose a sum lower than *sum* = 5 (or above *sum* = 9), you will be losing money on average. If you choose a sum between 6 and 8, you will be gaining money, and if you choose *sum* = 5 or *sum* = 9 you will, on average, end up at parity.

The Main Course: Hierarchy of Events

After enjoying the aperitif, you are now ready for the main course. In the two previous examples, we dealt with *events* (the outcomes of throwing dice) and a measure assigned to those events — their probabilities. We shall now explore in more detail the relationship between different *kinds* of events and their assigned probabilities (if such probabilities can be assigned at all).

Let us start with a very simple event: The occurrence of the result "4" in throwing a fair die. The assigned measure to this event is its probability. If you throw a fair die — a die that you have no reason to suspect is not homogenous or has any distorted shape — you *know* that the chance of obtaining each outcome is 1 out of 6; that is, the probability of the specific result "4" is 1/6.

Let us group all the games that have a finite number of outcomes (say, throwing a die, or two dice, or a coin, or 10 coins), and assume that all the outcomes have *equal probability*. The single events are the analogs of the atoms that we discussed in Section 2.1. The "weight" we assign to each event in this case is its probability. Thus, the "size" of the event — the outcome of throwing a die is "4" — is 1/6. I should add that we are focusing on the concept of *probability*. It is presumed that we know what an *event* is. The reader should be aware of the fact that the seemingly innocent concept of "event" might be ambiguous. Everyone knows what occurred in

New York on September 11, 2001. Was that a single event or two distinct events? This is more than a mere question of semantics.[3]

Back to probabilities, we define a dim-event as a collection of *specific* events having a common property; for instance, the event "the outcome of throwing a die is an even number." That means the outcome is either a "2," a "4," or a "6." The "size" of this event is the sum of the probabilities of the three *specific* events; that is, $1/6 + 1/6 + 1/6 = 3/6 = 1/2$. Thus, the probability of a dim-event is the analog of the weight of an object made of pure element — it is the sum of the probabilities of the "atomic" events.

We also assign to the event "any of the possible outcomes of the game has occurred" the "size" of unity. That means that if we throw a die, one of the possible outcomes has surely occurred. We call this event the *certain event*, and we assign to it the measure of unity.

We shall refer to the games (or experiments) having elementary outcomes of equal probability, as uniformly distributed games. The collection of all possible uniformly distributed games is denoted by *UDG* in Fig. 2.8.

Next, we expand the repertoire of games to include also non-uniformly distributed games. We still assume that we perform a game or an experiment and that there is a *finite* number of outcomes, but now the outcomes are not necessarily of equal probabilities.

A simple example of such a game is shown in Fig. 2.9a.

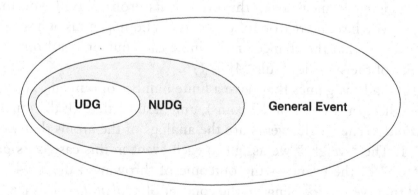

Fig. 2.8 Grouping of events.

We have 10 marbles in an urn, 1 red, 2 green, 3 yellow, and the remaining 4 are blue. Blindfolded, you picked up one marble at random from the urn. What is the probability that the marble you picked up is green?

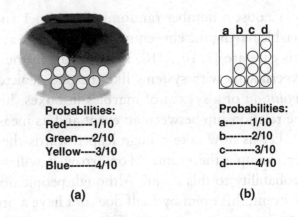

Probabilities:
Red-------1/10
Green----2/10
Yellow----3/10
Blue-------4/10

(a)

Probabilities:
a--------1/10
b--------2/10
c--------3/10
d--------4/10

(b)

You can surely calculate this probability. But, if you hesitate, let me help you. Clearly, since the marbles are identical except for their colors, the probability of picking up a *specific* marble is

Fig. 2.9 Non-uniform distribution. (a) Marbles of different colors in an urn. (b) Marbles in different cells.

1/10. There are 10 equally probable *specific* events. Therefore, from the point of view of *specific* marbles, this game belongs to the set of uniformly distributed games; that is, belonging to the set we denoted *UDG* in Fig. 2.8.

However, I did not ask you about the probability of a *specific* outcome but rather the probability of a specific color. This is not a specific event but a *dim-*event. Since there are two green marbles, each of which has a probability of 1/10 of being picked up, the probability of picking up *any* of the green marbles is 2/10. Likewise, the probability of picking up a yellow marble is 3/10, and a blue marble is 4/10. We shall call the list of numbers {1/10, 2/10, 3/10, 4/10} the *probability distribution* of this particular experiment. Thus, from the point of view of the dim-events this game is a non-uniformly distributed game.

Clearly, the set of all possible games (or experiments) having any distribution is larger than the set we denoted by *UDG* in Fig. 2.8. In Fig. 2.8, the yellow region represents the UDG. The red region represents NUDG. The union (or the sum) of these two regions represents games having *any* well-defined distribution.

Let me describe an equivalent game with 10 marbles in 4 boxes denoted a, b, c, and d in Fig. 2.9b. The marbles are numbered from 1–10 as in the previous game, but now they are of the same color. A specific distribution of the marbles in the boxes is shown in Fig. 2.9b.

Choose a number randomly between 1 and 10. What is the probability that the marble having the chosen number is in box a, b, c, or d? Clearly, the probabilities in this case are $\{1/10, 2/10, 3/10, 4/10\}$. In the next three chapters, we shall conduct experiments with systems like the one depicted in Fig. 2.9b. We shall focus on the *evolution* of a system of marbles in boxes. In this section we are interested only in the relationship between an event and its measure.

Let us now take a huge leap towards the most general set of events. Consider for example the event: "Tomorrow it will snow in Jerusalem." Can we assign a probability to this event? Although people do talk about the "probability" of such an event, this event by itself does not have a probability in the sense that we used this term in the previous cases. The reason for this is the following: In all the previously discussed cases, we have described an experiment or a game, and we assumed that we *know* all the possible outcomes and assumed that these outcomes have certain probabilities. In the case of the event "Tomorrow it will snow in Jerusalem," it is not clear at all what the experiment is and what its outcomes are. If you ask someone who has never been to Jerusalem or visited Jerusalem even for a short time, he or she would have no idea what the chances are that it will snow in Jerusalem tomorrow. However, if someone had collected statistics about the weather in Jerusalem for many years, and with the assumption that the weather trends will be consistent in the future, then he or she could "predict" the weather in Jerusalem tomorrow, and also give a numerical estimate, usually as a percentage, of the *chances* that it will snow tomorrow. What that probability basically means is the extent of one's belief in the occurrence of that event, based on whatever information one has on the weather in Jerusalem.

As a second example, consider the following event: "The book you are holding was written by Shakespeare." Again, it is not clear to which experiment this event belongs, and whether or not one can assign probability to this event. Surely, it is not uncommon to hear people talking of the probability of such an event. Once again, the meaning of this kind of assigned probability is basically a measure of one's *belief* in the occurrence of the event based on whatever information is available.

Let me reiterate that each point in the schematic diagram in Fig. 2.8 is an *event* (or an outcome of an experiment). In the innermost region denoted *UDG*, each event is one result belonging to a game, the outcomes of which are uniformly distributed.

In the region denoted *NUDG*, again each point is an event but now this event belongs to a game, the outcomes of which are not necessarily uniformly distributed; that is, each outcome of the game might have different probabilities, as in the example in Fig. 2.9a and Fig. 2.9b.

In the outermost region, each point also represents something that we recognize as an event but in general it is not clear whether or not that event belongs to any game or to an experiment with well defined probabilities. These kinds of probabilities are highly subjective, sometimes completely arbitrary, and most often meaningless. Clearly, we cannot deal with such events within the framework of science, and therefore we shall discard these cases. The inclusion of the "general event" region in Fig. 2.8 is meant to provide a general orientation on where "our" events are located.

The important lesson we have learned from the discussion above and from Fig. 2.8 is that science starts with a very general concept — in this case, events. One then tries to find some kind of a measure which may be assigned to these events. If that is impossible, then search for a reduced set of events on which a measure can be found; in this case, the measure is the probability. As for our specific needs in this book, we narrow down further the range of events to only those events the probabilities of which *you* already know.

This process of "distillation" that we did is the same as that which we had done previously regarding the weights of physical objects. It will also be the same as in the case of the concept of *information* discussed in the next section.

The Dessert: Children's Perception of Probability

Following the original investigations by Piaget and Inhelder (1951),[4] many experiments were performed to determine when and how young children make decisions based on probabilistic reasoning.[5] The general conclusion reached by psychologists is that young children develop an intuition for the concept of probability, probably as a result of the fact that, in an early stage of their life, young children encounter an environment abounding with randomly occurring events.

In this dessert, I offer you a special and a delicious sample of experiments carried out by Falk, Falk, and Levin (1980).[6] I shall describe only a small fraction of their experiments; just enough to convince you that if young children understand, even though they cannot define probability, you can understand, too! I also hope that this

Fig. 2.10   Marbles in an urn, a very easy problem.

dessert will whet your appetite to look forward to the meal that awaits you in the next section, which includes another beautiful dessert.

Children aged 4–11 were presented with two urns and various numbers of blue and yellow marbles (Fig. 2.10). Each child was informed about the exact number of blue and yellow marbles put into each of the urns. One color, say blue, was chosen as the "pay-off color" or the winning color. The child had to choose between the two urns and randomly draw one marble from it. If s/he drew the winning color, s/he was rewarded with a piece of candy. If s/he drew the non-winning color, in this case yellow, s/he got nothing. Each child played the same game several times.

Many sets of such games with different proportions and different numbers of marbles were prepared. Also, the colors of the marbles, the winning color, and the locations of the urns were changed to preclude the possibility of the child choosing according to, for example, his or her favorite color or preferred urn.

Recall that the child knows the exact content of each urn although this cannot be seen from the outside. One of the most surprising findings of these experiments was that the percentage of children of all ages choosing the correct urn (i.e. the urn in which the probability of winning is larger) was quite high. A closer examination of the results of these experiments showed that the most common error, especially of the youngest children, was that they chose an urn not according to the relative probabilities but according to the *absolute number* of the winning color of marbles. For instance, in the game denoted "very easy game" in Fig. 2.10, most children

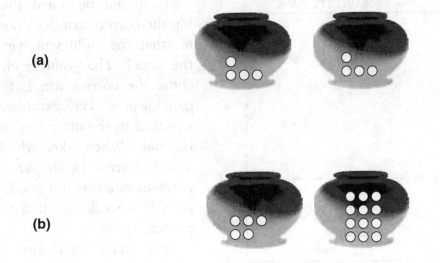

Fig. 2.11   (a) Easy and (b) difficult games of marbles in urns.

correctly chose the left urn. When asked why they chose that urn, their reply was: "Because there are more blue marbles in the left urn."

To eliminate the possibility of choosing the correct urn but for the wrong reasons, two sets of games were designed. In the easy games (Fig. 2.11a), the number of winning colors was larger in the correct urn (i.e. where the probability of winning is larger). In the difficult game (Fig. 2.11b), the number of winning colors was larger in the losing urn. Clearly, in the easy game the child can effortlessly choose the correct urn, simply because of the larger absolute number of the winning color, whereas in the difficult game the child would choose the wrong urn if he or she considers only the absolute number of the winning colors.

Fig. 2.12 shows the results obtained from the two games in Fig. 2.11. In both games there is a systematic improvement with age. In the easy game, both younger and older children scored well above the 50% level. On the other hand, with the difficult game, children aged below 6 scored poorly below the 50% level. However, there is a clear and abrupt improvement at around ages 6–7.

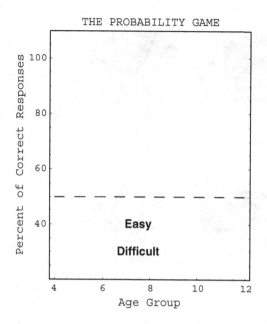

Fig. 2.12   Results from Falk, Falk, and Levin (1980).

It should be noted that choosing the correct urn does not guarantee that the child will win (i.e. get the prize). The younger child often chose the correct urn but failed to gain the prize. The next time, the child switched to the other urn (the incorrect one). When asked why he or she switched urns, the answer was: "The previous one was not good," — that is, it fails to deliver him or her the expected prize.

The main conclusion of these experiments is that aged below 5–6, children do not have a sense of probability. They choose an urn either at random, according to the absolute number of the winning color, or any other unspecified reason, such as preferred color or a preferred urn. Sometimes they stick to the chosen urn when it delivers the prize, or switch to the other urn when it fails to deliver, regardless of the probabilities.

At age 6 and above, most children make the correct choice, even though they do not know what probability is. At the age of 10 and above, children make the correct choice most of the time, even in the difficult game.

2.3. Information and its Shannon Measure

Aperitif: Playing the 20-Questions Game

A warm-up game for this section is the well known parlor game of "20 questions" (20Q). This game was very popular on radio in the 1940s, and later on TV in the 1950s (Fig. 2.13). There are different variations of this game — one popular version is called "Animal, Vegetable, Mineral." We shall play the "Person" version for now. Later, we shall play a more "distilled" version of this game.

Fig. 2.13 The 20Q game. A person thinking of a person.

I choose a person, and you have to find out who I have chosen by asking binary questions only. A binary question is one which is answerable only with YES or NO. Suppose I chose a person, say Einstein, and you have to find out which person I chose. The original game limited the number of questions to 20, but in our version of the game we will not limit the number of questions. However, you have to pay me 1 US$ for each answer you receive. When you discover the person I chose, you will get 20 US$.

Clearly, you know what you do not know; the person I have chosen. It is also clear that you want to acquire that unknown information by asking the smallest number of questions. Otherwise, you will be spending more than 20 US$ to get the 20 US$ prize when you discover the person I chose. Therefore, you have to plan your *strategy* of asking questions cleverly.

Before we start the game you might be wondering why 20Q? Why not 25 questions, or 100 questions? The answer will probably surprise you. Twenty questions are more than enough for all practical purposes. By "practical purposes" I mean games in which I choose an object, a person, an animal, or whatever, from a pool of objects that we both know. Obviously, I cannot choose someone you have never heard of. Therefore, in practice we implicitly limit the range of persons to those we are both familiar with.

Suppose you play this game with me. How many persons do you think you could choose that you are sure that I also know? Perhaps 1000, 10,000, or 100,000. I bet you cannot find more than 100,000 names to choose from.

If you are convinced that the "size of the game" is not larger than 100,000, or even 1,000,000, then you will be surprised to learn that 20 questions are more than enough to win the game (i.e. to spend less than 20 US$ on questions to get the 20 US$ prize). To achieve that you only have to be smart! If you are smart enough to play this game correctly then you will also be able to understand the Second Law.

Let us go back to the 20Q game. Remember, I chose a person and have assured you that you have heard of that person, or that we have agreed on a group of persons, such as those in Fig. 2.14. You have to find out who I chose by asking binary questions. The critical question is how to question!

There are many possible strategies of asking questions. The simplest is to *guess* the name of the person you believe I have chosen. A possible list of such questions is shown in the first column of Table 2.2. Another method is to ask if that person has some physical features or belongs to the same group — say, "does the person have blue eyes?" A sample of such questions is shown in the second column in Table 2.2. Another strategy is similar to the previous one, but now you ask for some attribute that roughly divides all possible persons into two halves; for instance, is the person a male? A sample of such questions is shown in the third column of Table 2.2.

What strategy will you choose to play this game?

If you choose the first strategy, you might guess the right person on the first question and win. However, since there are so many possible persons to choose from, your winning on the first question, although *possible*, is extremely unlikely. Take note

Fig. 2.14 A collection of persons for the 20Q game.

also that the larger the group of persons that we agreed upon, the harder your task is. Guessing directly the name of the person would be an inefficient way of asking questions. The reason is that with each NO answer that you get, you have eliminated one person. This means you are still left with almost the same *missing information* that you began with (i.e. the original number of unknown persons minus one). As we shall see later on, very young children do indeed choose this strategy as they are apparently bent on receiving the prize and they feel that this is the only way a YES answer will terminate the game in their favor and thereby win the prize.

However, if you choose the second strategy, you cannot possibly win on the first question. If you ask, "Is the person living in Paris?" and you get an affirmative answer, you still have to continue asking questions until you establish the identity of the person. If you get a NO answer, which is more probable, you exclude all persons

Table 2.2.   Different Strategies of Asking Questions

Dumb strategy: Specific questions	Smart strategy: grouping according to some property	Smarter strategy: grouping into two parts nearly half of the range of persons
1. Is it Bush?	1. Does the person have blue eyes?	1. Is the person alive?
2. Is it Gandhi?	2. Is the person living in Paris?	2. Is the person a male?
3. Is it Mozart?	3. Is the person an actor?	3. Does the person live in Europe?
4. Is it Socrates?	4. Does the person work in the field of thermodynamics?	4. Is the person in the sciences?
5. Is it Niels Bohr?	5. Is the person a male?	5. Is the person well known?

who live in Paris. This is certainly much better than excluding only one person, as done in the first strategy.

At this stage you feel that the best strategy is to divide all possible persons roughly into two groups. Again, you cannot win on the first question, or on the second, or the third. However, at each step of the game, whatever answer you get, whether it be a YES or a NO answer, you exclude a very large number of persons and you narrow your range of possibilities into almost half of the original number of possibilities.

Intuitively, you feel that in choosing the "smarter strategy" you gain more *information* from each answer you ask. Therefore, you get the maximum information for each dollar that you spend. It pays to be patient and choose the best strategy rather than rushing impatiently to guess the right person.

All I have said above is quite qualitative, but it is in the "right direction." For instance, you will agree that asking whether the person lives in Europe is better than asking if he or she lives in Paris, or London, or New York. Although I dubbed the third column in Table 2.2 as the *smarter* strategy, there exists no *smartest* strategy in this case. It is hard to find a strategy where on each question you divide the range of possibilities into exactly two halves. The best that you can do is to choose as nearly as possible to halve divisions, such as, "Is the person a male?" or "Is the person alive?" and so on. We shall soon see that in the "distilled" 20Q game there are cases for which a *smartest* strategy exists.

Exercise E2.2: Look at Fig. 2.14. There are some pictures of famous persons and some not so famous. Suppose we agree that I choose a person only from this collection. Is there a smartest strategy in this case? What is the minimum number of questions that you need to *guarantee* that you will get the required information?

There are also many subjective and psychological elements that enter into the general 20Q game. If you know me, you might have a rough idea of which person I am most likely to choose, and you will certainly be in a better position than someone who does not know me. Also, if I know you, and I know that you know me, and I know that you might think that I will be choosing person X, then I can outsmart you by deliberately choosing someone other than X. But then if you know me, and you know that I know you, and you know that I will try to outsmart you by choosing someone other than X, then you might "out-outsmart" me by betting on someone other than X. This kind of reasoning can drag on indefinitely.

Remember, this book is not about the 20Q game itself but about a *measure* of the game. Roughly, the size of the 20Q game can be measured by the amount of money you will spend in asking questions. Qualitatively, the larger the pool of persons from which we agree to choose the person, the larger the size of the game. If we agree to select only a person from the city in which we both live, then the size of the game is smaller compared with the game where we agree to choose a person from a country. If we agree to choose only persons that are in the room where we are attending a party, then the size of the game is even smaller.

Now that you know how to play the 20Q game correctly, you have implicitly admitted that you know what information is, and that this information has some measure or size. The information in this game is "which person I have chosen." You also know that if there are more persons from where I chose the one for the game, it will be *harder* to find that person.

Having enjoyed the aperitif (I presume), you should now be ready for some in-depth analyses of this game. We shall proceed along the same path with almost the same methodology we used in Section 2.2, to "distill" a simple concept of information to which we shall assign a measure, much as we have assigned a measure to an event (probability) and to a physical object (weight).

The Main Course: Hierarchy of Information

The concept of information is a very general concept. It encompasses anything we see, hear, or perceive with our senses. It includes subjective ("this book is interesting") and objective ("this book contains 200 pages") information. Information can be interesting or dull, it can be meaningful or meaningless, it can be helpful, beautiful, reasonable — add any adjective that pops up in your mind and you will find examples of such information.

The idea that information is something *measurable* was not widely appreciated until 1948, when Norbert Wiener's book, *Cybernetics*, appeared, and Claude E Shannon published *The Mathematical Theory of Communication*.

One can think of many measures of information, much like there are many measures of physical objects (weight, volume, surface-area, etc.) and of events (probability, length of the event, the number of objects the event is concerned with, etc.).

Consider the two following items of information that I have heard in the news today:

> "The snow storm that hit New York this morning has left a trail of devastation with 60 persons injured, 4 dead and thousands left with no electricity."
> "A bomb in Iraq killed 80 persons."

Clearly, the first is longer than the second. It has more letters and more words. It is also about more persons affected by the snow storm in New York. On the other hand, the second is shorter in length but reports on a *larger* number of persons killed. The second might also be more important and relevant for those who live in Iraq, but the first is more important for those who live in New York.

As you surely realize, it is even more difficult to assign a measure to *any* information, than it is to assign weight to *any* object, or to assign probability to *any* event. Therefore, we have to find a "distilled" form of information on which we can define a precise and objective measure.

As we have done in Section 2.2, we shall describe here two "families" of games, one based on uniformly distributed events and the other comprising of non-uniformly distributed events. The division into two families of games is formally the same as the two families of events discussed in Section 2.2. However, conceptually they are quite different. In Section 2.2 we were interested in the *events* themselves and their assigned measure. Here, we are interested in the *entire game* and its assigned measure. It just happens that the games also comprise events which might be uniformly or non-uniformly distributed.

We shall start with the simplest game and gradually upgrade the level of the game. The simplest example is a simplified version of the 20Q game. Instead of choosing a person from an unspecified number of persons, we have eight boxes (Fig. 2.15). I hide a coin in one of the boxes and you have to find where I have hidden the coin by asking binary questions.

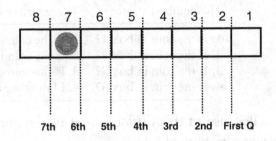

(a) The dumbest strategy

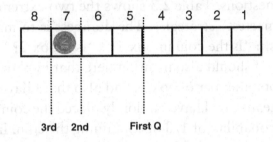

(b) The smartest strategy

Fig. 2.15 Eight boxes and a coin, the dumbest and the smartest strategies.

In this game, the *information* we are seeking is "where the coin is." What we are interested in is not the information itself but in some measure of the *size* of the missing information. Before we choose a measure of the size of the missing information, we note that the more boxes there are the more difficult it will be to find the missing information. What is meant by "more difficult" is that we need more questions to ask in order to obtain the missing information. One way to measure the size of the game is simply by the number of boxes (we will call this *NB*). Clearly, the larger the *NB*, the "bigger" the problem and the more questions we shall need to ask. That measure is fine, but it will be difficult to generalize for the case of non-uniformly distributed games which we shall soon discuss.

Another measure we can adopt is the number of questions we have to ask to find where the coin is. However, there is a difficulty with this measure. We already know that the number of questions depends on the *strategy* we choose in asking questions.

Table 2.3.   Different Strategies of Asking Questions

The dumbest strategy	The smartest strategy
1. Is the coin in box 1?	1. Is the coin in the right half (of the eight)?
2. Is the coin in box 2?	2. Is the coin in the left half (of the remaining four)?
3. Is the coin in box 3?	3. Is the coin in the right half (of the remaining two)?
4. Is the coin in box 4?	4. I know the answer!

Because of this difficulty, let us devote some time to playing this game with different numbers of boxes.

We start with eight boxes ($NB = 8$). We can choose many strategies in asking questions. Table 2.3 shows the two extremes: We call them "the dumbest" and "the smartest" strategies. The dumbest is to make a specific guess where the coin is and ask: "Is the coin in box 1?" "Is it box 2?" and so on.

I should also mention here that in playing this game, you are informed about the total number of boxes, and also that I have no preference for any specific box, which means that I have randomly placed the coin in one of the boxes; that is, each box has a probability of 1/8 of containing the coin. In the previous 20Q game we did not mention this explicitly, but it is important to do so in this game. Note that in this game we have completely removed any traces of subjectivity. You cannot use any information you might have about me or about my personality to help you in finding or guessing where I have placed the coin. The information we need here is "where the coin is." The "hiding" of the coin can be done by a computer which chooses a box at random.

Clearly, the thing you need is *information* as to "where the coin is." To acquire this information you are allowed to ask only binary questions. Instead of an indefinite number of persons in the previous game, we have only eight possibilities. More importantly, these eight possibilities each have equal probability of 1/8.

The strategies here are well defined and precise, whereas in the previous game I could not define them precisely. In this game, with the dumbest strategy you ask, "Is the coin in box k?", where k runs from one to eight. The smartest strategy is different: Each time, we divide the entire range of possibilities into two halves. You can now see why we could not define the smartest strategy in the previous game. There, it was not clear what *all* the possibilities were, and even less clear whether division by half was possible.

Second, note that in this case I use the adjectives "dumbest" and "smartest" strategies. (I could not do that in the previous game so I just wrote "dumb," "smart," and "smarter.") The reason is that here one can *prove* mathematically that if you choose the *smartest* strategy and play the game many, many times, you will out-perform any other possible strategy, including the worst one, denoted the "dumb-est." Since we cannot use the tools of mathematical proof, let me try to convince you why the "smartest" strategy is far better than the "dumbest" one (and you can also "prove" for yourself by playing this game with a friend or against a computer[7]).

Qualitatively, if you choose the "dumbest" strategy, you might hit upon the right box on the first question. But this could happen with a probability of 1/8 and you could fail with a probability of 7/8. Presuming you failed on the first question (which is more likely and far more likely with a larger number of boxes), you will have a chance of a right hit with a probability of 1/7 and to miss with a probability of 6/7, and so on. If you miss on six questions, after the seventh question you will *know* the answer; that is, you will have the information as to where the coin is. If, on the other hand, you choose the "smartest" strategy, you will certainly *fail* on the first question. You will also fail on the second question but you are *guaranteed* to have the required information on the third question.

The qualitative reason for preferring the smartest strategy is the same as in the previous game (but now can be made more precise and quantitative). By asking, "Is the coin in box 1?" you might win on the first question but with very low probability. If you fail after the first question, you have eliminated only the first box and decreased slightly the number of remaining possibilities, from eight to seven. On the other hand, with the smartest strategy the first question eliminates *half* of the possibilities, leaving only four possibilities. The second question eliminates another half, leaving only two, and in the third question, you get the information! We can also say that with the smartest strategy we reduce the "size" of the game each time by half.

In information theory, the amount of missing information — the amount of information one needs to acquire by asking questions — is *defined* in terms of the probability distribution.[8] In this example, the probability distribution is simple: $\{1/8, 1/8, 1/8, 1/8, 1/8, 1/8, 1/8, 1/8\}$. In asking the smartest question, one gains from each answer the maximum possible information (this is referred to as one bit of information). One can prove that maximum information is obtained by each question when you divide the space of all possible outcomes into two *equally probable* parts.[8]

Clearly, if at each step of the smartest strategy I gain *maximum* information, then I will get the information I want in a *minimum* number of questions. Again, we stress that this is true *on average*; that is, if we play the same game many, many times, the smartest strategy provides us with a method of obtaining the required information with the smallest number of questions.

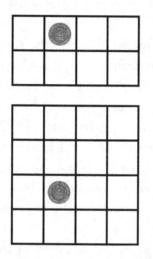

Note also that the *amount* of information that is required is fixed for a given game, and it is independent of the strategy you choose. The choice of the strategy allows you to get the same amount of information with a different number of questions. The smartest strategy guarantees that you will get it, on average, with the minimum number of questions.

If that argument did not convince you, try to think of the same game with 16 boxes (Fig. 2.16). I have doubled the number of boxes but the number of questions one needs to ask in the smartest strategy has increased by only one! The average number that one needs to ask in the dumbest strategy is far larger.

Fig. 2.16   Eight and 16 boxes.

If you are still unconvinced, think of 1,048,576 boxes. Using the smartest strategy, you are guaranteed to find the required information in 20 questions! Can you imagine how many questions you will need, on average, in the dumbest strategy?

The important point to be noted at this stage is that the larger the number of boxes, the greater the amount of the missing information, hence the greater the number of questions needed to acquire that information. This is clear intuitively. The amount of information is *determined* by the distribution (which in our case is $\{1/NB, \ldots, 1/NB\}$, for NB equally probable boxes).

If you use the dumbest method, with larger NB the average number of questions increases *linearly* with NB. This means that the average number of questions is proportional to the number of boxes. One can show that for very large NB, the average number of questions is about $NB/2$ (Fig. 2.17).

On the other hand, if you use the smartest strategy you will need only $\log_2 NB$ questions on average. Why $\log_2 NB$?

This follows from the property of the logarithm function (see Section 2.1). If we double NB, we get $\log_2 (2 \times NB) = \log_2 NB + \log_2 2 = \log_2 (NB) + 1$.

We have already seen that if we *double* the number of boxes, the number of (smart) questions increases by only one! This is a very important observation. This is also the reason why we shall adopt $\log_2 NB$ as a *measure* of the size of the game, rather than NB itself.

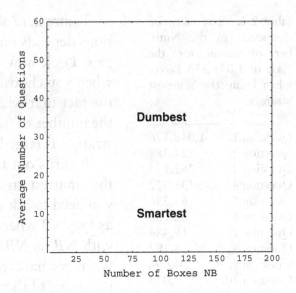

Fig. 2.17 Linear and logarithmic dependence on the number of boxes.

Now let us play the 20Q game with money. Suppose that we have 16 boxes, and you pay 1 US$ for each question that you ask. When you find where the coin is, you get 5 US$. Surely you would not like to ask direct and specific questions. If you do so, you will have to ask about eight questions on average, which means you will be paying 8 US$ to earn 5 US$. However, if you choose the smartest strategy then you will be spending only 4 US$ to earn 5 US$.

If this game did not impress you, think of having 1,000,000 boxes. You pay 1 US$ per question but you gain 1000 US$ once you find the coin. You would not even dare to use the dumbest method in this case, since you will lose on average about 500,000 US$! However, astonishing as it may sound to you, with the smartest strategy you will need to spend *less* than 20 US$ to gain the amount of 1000 US$.

If you are astonished by this, do not be intimidated. In fact, with 20 questions you can find the coin not in 1,000,000 boxes but in 1,048,576 boxes. If you do not believe me, try it out for yourself (see Table 2.4). It is very simple. Each time, divide the total number of boxes into two halves — in 20 questions you are guaranteed to find the coin $\left(\log_2 1,048,576 = \log_2 2^{20} = 20\right)$.

Table 2.4. The Stepwise Reduction in the Number of Boxes for the Case of 1,048,576 Boxes when Using the Smartest Strategy.

Question 1.	1,048,576
Question 2.	524,288
Question 3.	262,144
Question 4.	131,072
Question 5.	65,536
Question 6.	32,768
Question 7.	16,384
Question 8.	8,192
Question 9.	4,096
Question 10.	2,048
Question 11.	1,024
Question 12.	512
Question 13.	256
Question 14.	128
Question 15.	64
Question 16.	32
Question 17.	16
Question 18.	8
Question 19.	4
Question 20.	2

Figure 2.17 shows how the average number of questions depends on the number of boxes in the two strategies. Look how fast the number of questions increases when you choose the dumbest strategy. Be impressed by the fact that when the number of boxes becomes *huge*, the number of questions you need to ask in the smartest strategy is large but still manageable.

It turns out that for this particular game, if you use the smartest strategy the average number of questions you need to ask in order to find the coin depends on NB as $\log_2 NB$ whereas with the dumbest strategy it grows, with NB as $NB/2$. This is a huge difference, especially when we have very big numbers, and big numbers are the staple of thermodynamics!

The number of molecules in a quarter of a glass of water is about 10^{23}. If you have this number of boxes you will almost never be able to find the coin with the dumbest strategy. But with the smartest strategy you are guaranteed to find the coin with fewer than 80 questions!

Let us adopt the logarithm to the base 2 of the number of boxes as a measure of the size of the missing information in the game of the type discussed above. We shall refer to this measure as the Shannon Measure of Information and use the abbreviation *SMI*. If you are not comfortable with the concept of logarithm, forget it. Just remember that the smartest strategy of asking questions is indeed *the smartest strategy*, and take the average number of questions one needs to ask in the smartest strategy (*ANOQONTAITSS*) as a *measure* of the size of the missing information.[9] You can use the abbreviation *ANOQONTAITSS* or its synonym, *SMI*, which I prefer. The latter will accompany us throughout the entire book.

You should also realize that the type of information for which the *SMI* is applicable is but a tiny fraction of all possible types of information. We cannot apply this measure for most types of information. For instance, the probability that "a snow storm is expected at 3 p.m. tomorrow in the New York area" cannot be measured by the *SMI*. However, information theory can deal with the *size* of this message, regardless of its meaning or the information it conveys. This type of *SMI* is important in the field of communication and transmission of information, for which Shannon constructed his measure.

Having dealt with the simple games of a coin hidden in *NB* boxes, let us now discuss a more complex case. It is relevant to both communication theory and to information theory, but we shall deal with it from the point of view of the 20Q game.

Suppose I am thinking of a four-letter word (not what you are thinking of!). Again, you have to find out what the word is by asking binary questions. You know that the word is in English, and you know English. You also have some idea about the relative frequencies of the various letters in the English language (Fig. 2.18).

There are 26 possible letters, therefore there are 26^4 possible "words" of 4 letters. This is quite a large number of words — 456,976 (or about $\log_2 456976 \approx 19$). This means that if we know nothing about the English language we need to ask about five questions ($\log_2 26 \approx 5$) to find out which is the first letter (1 letter in 26 boxes), another five for the second letter, five for the third, and five for the fourth. Altogether, you will need about 20 questions. (It is not exactly 20 questions because you cannot divide into two halves each time.)

However, if you know something about the distribution of letters in English and you know English, you can do much better. Instead of dividing the total number of letters into two halves, you can divide into two parts of equal probabilities; for instance, the first 6 letters have the total probability of about 0.5, and the remaining 20 letters also about 0.5. The idea is that we give more weight to letters that appear more frequently than others. Based on the knowledge of the distribution of letters as provided in Fig. 2.18b, one can calculate that the *ANOQONTAITSS* or the *SMI* is about 4.[10] This is certainly better than 5, when you do not know the frequencies. It is easy to divide each time by "half" if we plot the distribution of letters in a pie-chart, as shown in Fig. 2.18c. The numbers correspond to the letters sorted according to their frequencies in Fig. 2.18b.

Once you find the first letter — say it is the letter *T*, you can proceed to the next letter. Again, if you know the frequencies of the letters you can find the second one by asking about four questions. However, if you know the frequencies of *pairs* of letters, you might use this information to reduce the number of questions needed to

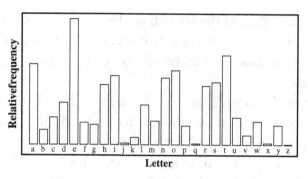

(a) In alphabetical order.

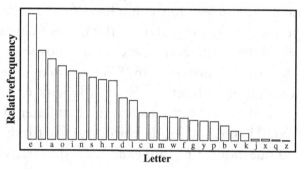

(b) In the decreasing order of relative frequency.

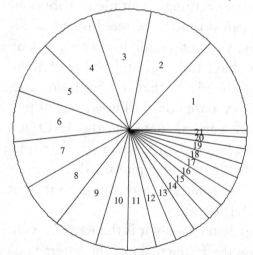

(c) As (b) but the numbers indicate the letters in the same order as in (b).

Fig. 2.18   Relative frequencies of letters in English.

find the second letter. For instance, the combination *th* is more frequent than the combination *tl* and certainly more frequent than the combination *tq*. Therefore, you can use this information to reduce the number of questions, let us say to about three (or perhaps fewer).

Suppose that you found the first two letters are *th*. Next, you need to find the third letter. That is relatively easy. It is most likely that the third letter after *th* is a vowel — *a*, *e*, *i*, *o*, *u* — so you have to find one of five letters, for which, on average, you will need about two questions (note that *e* and *a* are more frequent than *i*, *o*, or *u*). Suppose you did that and found that it is an *i*; you therefore already have three letters, *thi*. Now you do not need any more frequencies. If you know the English language you know that there are only two words that are four lettered and start with *thi*; *this* and *thin*. So with one more question you will find the word I had chosen.

The important thing to understand is that whenever you know something about the distribution of letters (or the boxes, or the persons from which I selected one) you can use this information to *reduce* the number of questions you need to ask. In other words, any information you have on the game reduces the *size* of the game or the size of the missing information of the game. We shall soon discuss a more distilled game of this kind, but let me first tell you a real story of how I could use information to reduce the missing information. As a young man, I served in the army. We took turns in patrolling the camp for a few hours during the night. To pass the time we played the 20Q game. My partner was quite proficient in English as he had been educated in America. Of course, he had an advantage over me. I chose a four-letter word and in only eight questions he guessed the word. Then we reversed roles. He chose a four-letter word, and I had to find out by asking binary questions. He was stunned when I guessed the word in only two questions! "How did you do it?" he asked. "That is simple," I answered. "I know you, and I know what always goes on in your mind, and I used this information to my advantage."

Figure 2.19 shows three nested regions. The two inner regions are exactly the same as the ones in Fig. 2.8. The only difference is that here we collect events belonging to the same game (or experiment) in circles. The blue ellipse is reserved for the entropy and is discussed in Chapter 7.

The innermost region, denoted *UDG* (uniformly distributed games), includes all the games of the type of a "coin hidden in *NB* boxes"; that is, games with a finite

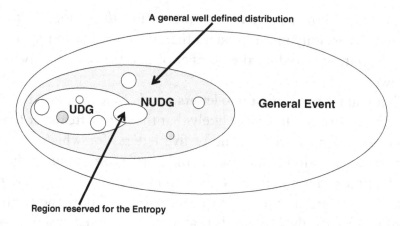

Fig. 2.19 Nested regions of events. Small circles include events belonging to the same game or experiment. The dark blue ellipse is discussed in Chapter 7.

number of outcomes and where each outcome is equally likely to occur. As in Fig. 2.8, the points in this figure represent a single "atom" of events — "the coin is hidden in box 6 in a game with 8 boxes," or "the coin is hidden in box 26 in a game of 32 boxes." Unlike in Fig. 2.8, where we assigned a measure to each single event, here we assign a measure to the whole game. These are shown as small colored circles in Fig. 2.19. Thus, the measure is not assigned to each *point* in the figure but to each *game*, or to each circle in Fig. 2.19. A bigger game has a bigger size, or a bigger area in Fig. 2.19. Note the region marked in dark blue in Fig. 2.19. This region represents the entropy which is a special case of SMI. See also Chapter 7.

Thus, the content of the sets in Fig. 2.19 is similar to that in Fig. 2.8 but now we assign a measure to each game, which in this particular game could be chosen to be the number of boxes. However, as I have explained above, it is advantageous to adopt the logarithm to the base 2 of the number of boxes, rather than the number of boxes itself. In this particular case, this seems to be only an incidental technicality. However, when we generalize our games to include the non-uniform distributed game, we shall see that this technical difference is of paramount significance. For this reason we adopt here the logarithm to the base 2 of the number of boxes as a measure of the size of the game. In practice, we shall be using a simpler but equivalent measure, which is easier to grasp; namely, we take the *ANOQONTAITSS*, or simply the *SMI*, which in this game is also equal to the logarithm to the base 2 of NB.

The reason I am switching to this new measure is twofold; first, it is simple and easier to grasp, and second, it can also be applied to the more general game without modification. The original measure discovered by Shannon is mathematically more demanding. The equivalent measure (*ANOQONTAITSS*) as stated above does not need any mathematics; simply play the game and count the number of questions you ask, provided you are smart enough not only to *count* but to choose the smartest strategy of asking questions.

I cannot resist informing you that you *are capable* of choosing the smartest strategy. If you have any doubt about that, go and taste the delicious dessert that awaits you at the end of this section.

Now for the generalization of the game. This part is the most profound, most important, and the most beautiful part of this book. It is not difficult, although it might appear to be so. Therefore, I shall be slow and use enough game terminology to make you feel comfortable.

Note again that the term non-uniform means here *any* distribution or a general distribution. A uniform distribution is a special of general distribution.

We shall describe two games that are generalizations of the equally distributed game discussed at the beginning of this section. The two games are equivalent. The first is easier to grasp, the second is formulated in terms of marbles in boxes, and requires a little more imagination. Playing the second game will facilitate the passage to the next three chapters, where we shall be playing only with marbles distributed in cells. More importantly, this game will also ease our passage to real systems, which we will discuss in Chapter 7.

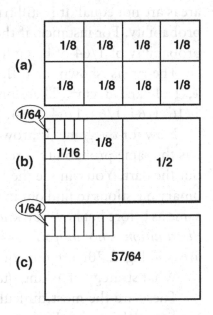

Fig. **2.20** Boards with different partitions.

A Dart Hitting the Board

Suppose that instead of hiding a coin in one of eight boxes, I throw a dart onto a dartboard while I am blindfolded. The board is divided into eight squares of equal areas (Fig. 2.20a). After hitting the board,

I look at the square in which the dart hit and ask you to find out this square by asking binary questions. Note that by throwing the dart with a blindfold on, I make sure that the dart hit somewhere in the entire board, and each of the squares of equal areas have the same probability of getting the dart. You realize that this is exactly the same problem as in the game of a coin hidden in one of the eight boxes, only here I made it clear and explicit that the eight areas have equal probabilities.

Exercise E2.3: You are given two boards, like the one in Fig. 2.20a, each of the same area and each divided into eight squares of equal area. I threw a dart blindfolded and it hit one of the two boards. How many questions do you need to ask to find out in which square the dart hit?

Exercise E2.4: I threw two darts, one on board 1 and the second on board 2. How many questions do you need to ask to find out in which squares the two darts hit?[11]

Next, we modify the structure of the board as shown in Fig. 2.20b and repeat the same procedure. Again, we have eight areas but the probabilities of hitting the different areas are not equal. It is still true that each unit of area — say, 1 cm^2 — has the same probability. For instance, if the total area is 64 cm^2, each square centimeter has the probability of 1/64 of being hit.

The areas shown in different colors in Fig. 2.20b are about 32, 16, 8, 4, 1, 1, 1, and 1 cm^2. Therefore, the corresponding probabilities are 1/2, 1/4, 1/8, 1/16, 1/64, 1/64, 1/64, 1/64, respectively.

Now for the game. I throw a dart while blindfolded. Therefore, each *unit* of area has the same probability. I remove the blindfold to see where the dart hit and pluck out the dart. You can see the board with its divisions in Fig. 2.20b. You have to ask binary questions to find out in which of the eight regions the dart hit. The game is the same as before but now you know, simply by looking at the board, that the *probability distribution* is *not uniform* as in the previous game. This simply means that the eight areas in Fig. 2.20b are not equally probable, as was the case in the Fig. 2.20a.

What strategy of asking questions will you choose?

These are the most difficult questions you are going to encounter in this book. Let me explain why. In the uniformly distributed game of eight boxes (Fig. 2.20a), you were convinced that asking the smartest strategy was to your advantage. In the specific game of eight boxes (Fig. 2.20a), you can be guaranteed that you will find the coin in three questions. Any other strategy would entail, on average, more questions.

Therefore, it is tempting to also use here the same *three* questions to get the missing information in the game in Fig. 2.20b. Indeed, you can certainly get the information in three questions. Simply divide the eight areas in any arbitrary two groups of regions each containing *four* regions (e.g. the four small regions marked by 1/64 and the other four). This way, after the first question you would already have reduced the number of possibilities from eight to four. Do the same again by dividing into two groups each containing two regions. This will reduce the number of remaining possibilities from four to two, and on the third question you get the required information. That sounds very efficient, right? Wrong!

Information theory teaches us that you can do better in this game. You can obtain the required information not from three questions, but on *average*, from fewer than three questions. The idea is the same as in the four-letter game. If you know the distribution and if the distribution is non-uniform, you can always *reduce* the number of questions. In other words, any information you have on the system can be used to your advantage. (Remember how I beat my friend, even though he knew English better than I did?)

The actual calculation of the average number of questions for these games requires some mathematics. I will give you the answer here.[12] It is 2 (not 3)! Here is a qualitative explanation.

The general principle of information theory is that whenever you have a non-uniformly distributed game, you can exploit the non-uniformity to your advantage: You can use the *information contained* in the distribution to get the required information faster.

Before discussing the meaning of the number 2 I quoted above, allow me to convince you of the veracity of the principle with a simpler example. Suppose that again you have *eight* regions as shown in Fig. 2.20c, but now the areas of the eight regions are 1, 1, 1, 1, 1, 1, 1, and 57 cm^2. The *total* area is again 64 and it is divided into eight, non-overlapping regions.

Surely you would not divide the eight regions into two groups of four, then divide into two groups of two, then get the required information on the third question. Why? Because the probability that the dart hitting the larger region is 57/64, whereas the probability of hitting one of the small regions is, yes, a slim probability of 1/64. Therefore, it will be wiser to ask first the question: "Did the dart hit the area marked 57/64?"

The probability of winning the game in *one* question is quite large. Of course, if you play the game several times you might not succeed in the first question. In such a case you will have to ask three more questions, so that altogether you will end up asking *four* questions. However, this case is rare. If you play many times you will, on average, ask 0.8 questions, or roughly one question only.

How did I get this number? I used Shannon's measure.[13] But a rough calculation is as follows. You have a chance of 57/64 of winning with one question, and you have a chance of 7/64 of winning with three questions. Therefore, the average number of questions is about:

$$\frac{57}{64} \times 1 + \frac{7}{64} \times 3 \approx 1.2$$

This is only an approximate average — the more exact value is 0.8; that is, on average you need about one question.[13]

Do the following exercise before returning to the game in Fig. 2.20b.

Exercise E2.5: A board is divided into 1000 regions. One region has an area of 999 cm^2 and there are 999 smaller regions with a combined area of 1 cm^2. How would you choose to play this game?[14]

We return to the game associated with Fig. 2.20b. As you have convinced yourself with Exercise E2.5, the smartest strategy is not the one when you divide the total *number* of possibilities into two halves. This was true in the uniformly distributed game. In the more general game, the smartest strategy is to divide the total number of possibilities into two parts of *equal* (or nearly equal) probabilities. Thus, for the game in Fig. 2.20b, the smartest strategy is to ask the questions according to the scheme depicted in Fig. 2.21. The first question: "Is it in the right half of the board?" There is a 1/2 chance that you will be out of the game at this stage. If not, continue to divide the remaining area into two equally probable regions and ask: "Is it on the upper part of the remaining board?" and so on. This looks like you are following the "dumbest" path in the game of Fig. 2.20a, but it is actually the smartest. In both

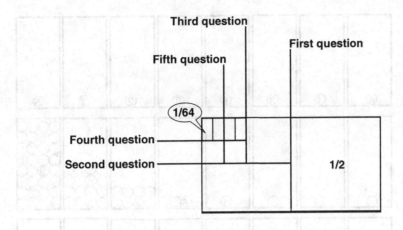

Fig. 2.21 The order of asking questions.

cases, the smartest strategy is not to divide into two *equal number* of possibilities but into two parts having *equal probabilities*.

One can prove mathematically that if you follow the smartest strategy you gain the *maximum amount* of *information* at each stage when you divide the total number of possibilities into two halves of equal probabilities. The amount of information obtained by each question translates into minimum number of questions. The units chosen to measure information are *bits* — short for *binary digit*. Asking a question about two equally probable possibilities is, in these units, one bit.

It is not always possible to divide into two equally probable regions. For instance, in the game in Fig. 2.20c there is no such division. In this case, divide as closely as possible to two equal probability possibilities.

This brings us to the next game. Mathematically, it is equivalent to a game of darts hitting a board. However, in practice it will be easier to do *experiments* with this game in the next three chapters.

The New Game of Marbles in Cells

Let us first translate the game of Fig. 2.20a into a language of marbles in cells. We have eight cells, and in each of them we place one marble. The marbles are numbered 1–8 and they are distributed at random in the eight cells, as shown in Fig. 2.22a. I choose a number between 1 and 8, and you have to find out in which cell the marble with the chosen number is.

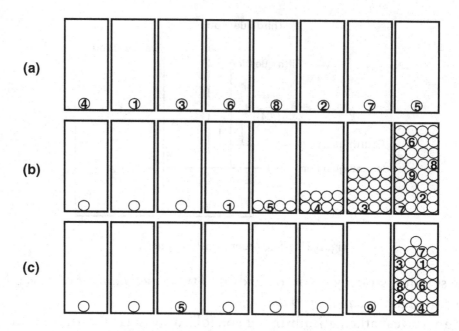

Fig. 2.22   Marbles in cells.

You should realize that the two games in Fig. 2.20a and Fig. 2.22a are completely equivalent. You have eight possible outcomes and the distribution of probabilities is $\{1/8, 1/8, 1/8, 1/8, 1/8, 1/8, 1/8, 1/8\}$, exactly as in the previous game.

The game in Fig. 2.22b is similar to the one in Fig. 2.20b. We have eight cells but the total number of marbles is now $32 + 16 + 8 + 4 + 4 = 64$. Again, the marbles are given numbers from 1 to 64 (only a few numbers are shown in the figure). The *numbers* are randomly distributed over the marbles. Note that the configurations shown in Fig. 2.22b and Fig. 2.22c are *specific* configurations; that is, we know which marble is in which cell. If we erase the numbers on the marbles, we get a dim-configuration; we shall only know how many marbles are in each cell.

In this game, I choose a number between 1 and 64, and you have to find out in which cell the marble with the number I chose is located. Clearly, the more marbles in a cell, the larger the probability that the marble marked with the number I chose will be in that cell.

The game in Fig. 2.22c is similar to the game in Fig. 2.20c. Try to play this game and estimate how many questions you will need, on average, to find where the marble I had chosen is located.

Exercise E2.6: Write the probability distribution for each of the games in Fig. 2.22. The probability of finding a specific marble in cell x is simply the number of marbles in cell x divided by the total number of marbles. The collection of the probabilities $(p_1, \ldots, p_n)$ is called the probability distribution for this game.

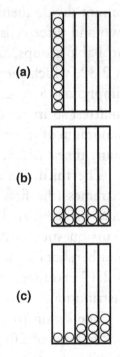

We shall refer to these distributions as the *state distribution*; these distributions describe the state or the configuration of the system. Each component of the distribution is the probability of finding a *specific* marble in a *specific* cell. In the next chapter we shall define a new probability, which will be referred to as the *probability* of the *state*. This is potentially confusing, therefore I urge you to make sure that you fully understand what the *state* probability distribution of the game is before we introduce the *probability* of the *state*.

Let us look at another example with a different state probability distribution. We will use this example to practice first, what a state distribution means, and second, to practice with a game in which we cannot divide all possibilities into two equally probable halves.

Consider the game depicted in Fig. 2.23a, in which we have 5 cells and 10 marbles all placed in one cell. What is the *state* probability distribution for this game?

Very simply, the state probability distribution is {1, 0, 0, 0, 0}. What this means is that there is probability 1 (i.e. certainty) that the chosen marble (numbered between

Fig. 2.23 Ten marbles in five cells.

1 and 10) will be found in the first box, and there is probability of 0 that the chosen marble will be in any other box.

How many questions do you need to ask in order to find out where the chosen marble is? None. We already know where the chosen marble is, whichever marble was chosen.

Next, consider the game depicted in Fig. 2.23b. Again, we have 10 marbles and 5 cells, but the *state* probability distribution is *uniform*: two marbles in each cell.

We play the same game, I choose a number between 1 and 10 and you have to find out in which cell the chosen marble is. The *state* probability distribution is simply $\{1/5, 1/5, 1/5, 1/5, 1/5\}$; that is, you are equally likely to find the chosen marble in any one of the cells. The probability of finding the marble in the first cell is $1/5$, and so is the probability in any other cell. How many questions do you need to ask in order to find out where the chosen marble is? We have five possibilities but we cannot divide them into two equal probability halves, so we divide into two parts; two and three cells first, and then divide again into, as far as possible, two equally probable groups, and so on. The exact Shannon measure of information in this case is 2.32, which means that on average you will need about 2.32 questions (this is simply $\log_2 5 = 2.32$). Qualitatively, you can convince yourself that by using the smartest strategy you would find the marble with a minimum of two questions and a maximum of three questions. The average number of questions, if you play the game many times, is 2.32.

The third example in Fig. 2.23c is somewhat intermediate between the two extremes (the first requires 0 questions, the second requires 2.32 questions). The *state* probability distribution for this case is $\{1/10, 1/10, 2/10, 3/10, 3/10\}$. How many questions will you need on average to play this game? It is difficult to guess this number, but calculation based on information theory shows that the average number is 2.17. You see, this is a little better than the number 2.32 in the case of uniform distribution.

The lesson from all these games can be summarized as follows. For each game of the kind of 20Q, with a finite number of outcomes and given *state* probability distribution, there is a quantity that characterizes the *entire* game and it is called *SMI*. I cannot tell you how Shannon originally defined the *SMI* without using mathematics. But I can tell you that this measure is the equivalent to the *average number* of *binary questions one needs to ask in the smartest strategy* (*ANOQONTAITSS*). This number is not only easy to understand but also easy to calculate, simply by playing the game.

Furthermore, if the distribution is such that all the marbles are in one cell — that is, the probability distribution is $\{1, 0, 0 \cdots\}$ — then you have to ask the minimum number of questions, which is zero! If, on the other hand, the distribution is uniform — that is, equal probability to each possibility — then the average number of questions is maximal. It is, in fact, simply the logarithm to the base 2 of the number of possibilities. Any other game with a distribution between these two extremes has a typical *SMI* which is approximately equal to the average number of questions you need to ask; this number falls between the minimum 0, and the maximum $\log_2 NB$.

Now that you know what *state distributions* are, let me just give you a brief explanation of what the *distribution* of *states* is.

In all of the games of marbles in cells that we played in this section, we have *defined* the game in terms of the dim-state; that is, how many marbles are in each cell. These numbers determine the *state* probability of *this game*. We can play the same game many times. The *state* distribution remains the same, simply because the game is fixed. In the next chapter, we shall "shake" the game: We shall let marbles jump from one cell to another, to create *new* games. We shall "shake" the systems with *fixed numbers* of marbles and *fixed numbers* of cells, but the *game* will be changed every time a marble jumps from one cell to another. In this process, we can ask: "What is the probability of obtaining a *specific game*, out of all possible games having the same number of marbles and the same number of cells?" This probability will be referred to the *probability* of the *state*, or the probability of the *game*. The collection of all the probabilities of the *states* will be the *distribution* of *states*.

The Easiest Game You Ever Played

You are shown three games as in Fig. 2.24. In each you have two cells, and you are told that there is one coin hidden in one of the cells. You do not know where the coin is hidden but you are told that the cell where the coin was placed was selected with a specific probability distribution. In plain and simple terms, you know that in:

Fig. 2.24 Three games with two boxes.

game (a) there is probability 1 of finding the coin in the first cell and 0 in the second;
game (b) there is probability 0.5 of finding the coin in the first cell and 0.5 in the second; and
game (c) there is probability 0.75 of finding the coin in the first cell and 0.25 in the second.

For each of the games you are given 100 guesses to find where the coin is. Every time you make a correct guess, I set aside a piece of candy (I know where the coin is but I will not tell you, even if your guess is correct, until after the end of the game).

Remember, I have prepared the three games and at each step I have placed the coin with the specific probabilities as shown in Fig. 2.24.

> Which cell will you choose in game (a) and how many pieces of candy do you expect to win after 100 guesses?
> Which cell will you choose in game (b) and how many pieces of candy do you expect to win after 100 guesses?
> Which cell will you choose in game (c) and how many pieces of candy do you expect to win after 100 guesses?

I am absolutely confident that you will know the answers. However, I urge you to imagine playing this game, and perhaps inventing some other games of a similar kind, and trying it on your friends or family members. One more question to think about: "Why did I bother to suggest such a simple exercise in the context of this section?"

This has been a long course. We have to leave some room for the very special dessert awaiting us. Before that, let us briefly summarize what we have learned in this section. We started with simple games in which we had a certain number of boxes, or possibilities. You were told that one possibility was chosen, but there was no preference or bias in choosing any particular possibility. You had to find which of the possibilities was chosen, by asking binary questions. Then we had to assign a measure or a *size* of the problem (roughly, how difficult it is to find the chosen possibility out of N possibilities). Intuitively, we feel that the larger N, the more difficult the task. But we also found that by *doubling* the number of possibilities from N to 2N, the size of the problem increases only by one; that is, doubling the number of possibilities from N to 2N requires, on average, asking one more question. This suggests that the size of the problem should behave as the logarithm of N. For convenience, we choose logarithm to the base 2. Thus, if the number of possibilities is doubled from, say, $N = 8$ to $N = 16$, the size of the problem changes from $\log_2 8 = 3$ to $\log_2 16 = 4$.

We then proceed to a more complicated game where the possibilities are not equally distributed. Again, you have the task of finding the *information* where the coin is hidden, or where the dart hit the board with a minimal number of questions. We saw that whenever the probability distribution is not uniform, we can use this knowledge to get the required information, on average, with fewer questions.

Again, we can measure the size of the problem by the number of questions one needs to ask to obtain the required information. These types of games are denoted *NUDG* in Fig. 2.19.

Beyond this region of *NUDG* we enter into the vast domain of general event or information. We shall not need to discuss this region in Fig. 2.19 but it is important to realize that the regions UDG and NUDG are only small subsets of the vast region of the general events, or of general information. We shall see in Chapter 7 that "entropy" is a special case of SMI, which may fall either in the UDG or the NUDG regions.

The outermost region in Fig. 2.19 is admittedly vague and not well defined. We include here information of all kinds. For instance: Tomorrow it will rain in the morning. Clearly, this is information, colloquially speaking, but it is not clear whether or not this information belongs to any "game." Let's leave that for now, as the dessert is being served.

The Dessert: Children Playing the 20Q Game

This is the last dessert for this chapter, and in my opinion it is the most delicious of all the desserts we have had so far. In Section 2.1, the dessert involved relatively simple tasks — children's ability to choose between two possibilities — in which it was quite straightforward to make the correct choice.

In Section 2.2, we encountered children having to decide between two possibilities of unequal probabilities. In this section, we are interested in the way children ask questions to obtain information. This task involves not only a feeling for probabilities but also planning or choosing the best strategy for obtaining the required information in the most efficient manner.

Many investigations were carried out on the way children play the 20Q game. We shall discuss only a few examples.

First, we discuss the work of Mosher and Hornsby (1966).[15] Two versions of the 20Q game were used. In one, the children were shown an array of common objects totaling 42 pictures. Figure 2.3 shows the original array of pictures used by Mosher and Hornsby. Figure 2.4 shows another array of pictures that I have prepared for this book. The children's task was to find out which one of the objects the experimenter had in mind. They could ask only binary questions. Here, we focus on the ability of the children to find out the required information with the minimal number of questions.

Ninety boys aged 6–11 and in grades 1–6 were shown 42 pictures of familiar objects (something like those in Fig. 2.4). Each child was first asked to identify the pictures, to ensure that he or she was familiar with them. The following instructions were given to each child:

> "Now, we're going to play some question-asking games. I'm thinking of one of these pictures, and your job is to find out which one it is that I have in mind. To do this you can ask any questions at all that I can answer by saying 'yes' or 'no,' but I can't give any other answer but 'yes' or 'no.' You can have as many questions as you need, but try to find out in as few questions as possible."

The questions asked by the children were classified into three groups. The first, called "hypothesis," where the child asked about a *specific* object, such as: "Is the object the dog?" This is what we have referred to as the dumbest strategy. The second, called "constraint seeking," is similar to what we have referred to as the smart strategy (not necessarily the smartest). In this strategy, the child divides the entire range of possibilities into two groups. The best strategy (the smartest) is to divide the entire range into two parts of equal number of objects each time. (Recall that this is the smartest strategy, provided that all the objects have the same probability. This is not part of the explanation given to the child; however, it is implicitly implied.) In practice, the children used some criterion for grouping, such as, "Is the object an animal?" or "Is it a food?" Clearly, this is better than the dumbest strategy but short of being the smartest.

The third strategy is referred to as "pseudo-constraint." This is essentially the same as the specific hypothesis, but it is phrased like a constraint question. For instance: "Does it have two wheels?" or "Does it bark?" Clearly, in this particular array of objects there is only one object that has two wheels, or barks. However, the question could be a "constraint" question in another game where there are different objects with wheels, or different kinds of dogs.

The results are shown in Fig. 2.25. As is expected, the first-graders (aged about 6) almost always chose the "specific hypothesis." Out of the 30 children, only 5 asked some constraint questions. From the third-graders (aged about 8), only about one-third asked "specific hypothesis" type of questions, and of the sixth-graders (aged about 11), almost all asked constraint questions. As can be seen from Fig. 2.25, the pseudo-constraint questions were used by fewer than 10% of the subjects, almost independently of their age.

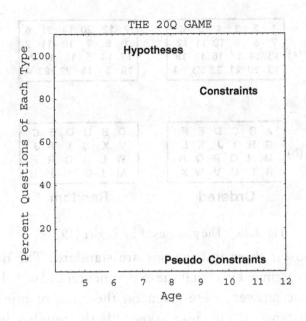

Fig. 2.25   Results from Mosher and Hornsby (1966).

Clearly, the choice of the constraint type of question requires some cognitive skills, planning, and patience. Some mature thinking is required to invest in questions that certainly cannot provide them with the required information immediately but that prove more efficient on average. Whereas young children leap hurriedly to ask specific questions, hoping to be instantly successful, the older ones invest in thinking and planning before asking.

Before concluding this section, I cannot resist the temptation of telling you of some further experiments on how adolescent children develop the skills of asking questions. I will make it very brief. It is also interesting to see how exactly the same two games can sometimes be perceived as having different degrees of difficulty.

Siegler (1977) has performed two sets of experiments with children aged 13–14.[16] The children were successively presented with two matrices, each with six rows and four columns. One matrix contained the numbers 1–24, while the other one contained letters A–X (Fig. 2.26). The procedure in these experiments differed from the standard 20Q game in one major respect. The subjects were informed about the

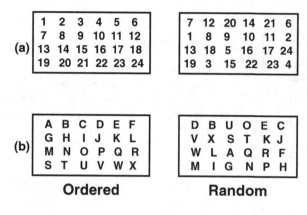

Fig. 2.26 The game used by Siegler (1977).

rules of the game, exactly as in the 20Q game; that is, they were given the following instructions:

"I am thinking of a number (or letter) from this matrix. You can ask questions that I can reply to with a YES or a NO answer. You have to find out which number (or letter) I had in mind in as few questions as possible."

So far these instructions are standard. The difference in the methodology of this experiment was that the experimenter did not choose any number or letter in advance. The answers were given on the basis of minimal elimination of alternatives. For instance, if the child asked, "Is the number between 1 and 10?" he or she would be told NO, because this answer would eliminate only 10 possibilities. If, on the other hand, the child asked, "Is the number between 1 and 20?" he or she would be given the answer YES, because this would eliminate 14 possibilities (out of 24). This procedure ensured that the child would never get a YES answer if he or she asked a *specific* question, such as, "Is the number 7?"

Clearly, the minimum number of questions required in this game using the smartest strategy is five (check this). However, if the child asked questions by guessing specific numbers (or letters), on average, he or she would require more questions than in the standard 20Q game.

Although the game with 24 *numbers* is equivalent to the game with 24 *letters*, the results obtained for the two games were different. Also, the results were different when the numbers (or letters) were arranged in the natural order, or randomly. See also Self-Testing kit in the Preface.

Interestingly, the results show that children did systematically better (i.e. fewer questions) on the number matrix (Fig. 2.26a) than on the letter matrix (Fig. 2.26b). Also, they did better when the numbers (or letters) were in their natural order than when the numbers (or letters) were placed at random. Can you think of an explanation for this result?

Finally, it is of interest to mention one further experiment by Marschark and Everhart (1999).[17] In this experiment, 36 hearing-impaired children and 36 children with no hearing disabilities, aged 7–14, participated. The game was essentially the same as the one used by Mosher and Hornsby (1966). It was found that in all the age ranges, those children with normal hearing capacity scored better than those who were hearing impaired (i.e. the former asked more constraint questions compared with the latter). Possible causes for these results were discussed in the article.

Here, we end our last dessert. I hope that this and the previous desserts have given you some "food for thought." Personally, I was fascinated to learn how young children develop, at quite an early stage of their lives, both a sense of probabilities and the skills in choosing an efficient strategy for playing the 20Q game. The purpose of presenting these results in this book is mainly to convince you that if you are above 11 years of age, there is no reason for you not to be able to understand the Second Law of Thermodynamics.

We now have to get to work. In the next three chapters, we shall carry out experiments on systems consisting of marbles in cells. I promise you that if you follow attentively the results of the experiments, you will be rewarded.

Snack: Newsworthy, News Crazy

I read in one local newspaper some very disturbing statistics: Apparently, 25% of Israeli women suffer from depression, while 17% of Israeli men suffer from the same fate. The reporter concluded that 42% of Israel's population suffers from depression. Does this alarming news mean that everyone in Israel should have a check-up?

On the lighter side, I read in another newspaper that the overall performance of Highiq State University's (HSU) students improved last year. The grades of *all* the students were significantly *above the average*. Isn't that good news, for a change?

In another publication, it was reported that in Highiq State University (HSU), the average IQ of university professors was 130. In that same publication, it was also reported that the average IQ of university professors at the Lowiq State University, (LSU) which is located in the next town, was only 80.

Two years ago, a professor from the HSU took a position in LSU. At the end of the academic year, it was published that as a result of this transfer of the professor, **the *average* IQ of the professors of each of the two universities had increased!**

A year ago, another professor from the HSU moved to LSU. At the end of that academic year, startling but otherwise welcome news were again published. As a result of this second transfer of the professor, **the *average* IQ of the professors of each of the universities had further increased!**

Can that be possible?

Yes, it is possible. You can easily find an example.

Does the increase of the average IQ of the professors of each university imply that the average IQ of *all the professors* in the two universities has also increased?

No, that is not possible. Think why?

Snack: Fifty-Percent Off or a Rip-Off?

A few years back I got in touch with a reputable real-estate agency and told them of my intentions to purchase a new apartment. A young and seemingly inexperienced sales agent took me to several nice apartments, one of which I really liked.

I had almost decided to buy the apartment when the agent handed me a contract which stated that the real estate company charges a 15% commission from the apartment's total value.

I thought that 15% was rather too high and so I tried to haggle with the agent. Since I was going to sell my old apartment anyway, I thought it might be a good bargaining chip. I told the lady that if she agreed to lower their commission I would sell my old apartment, and buy a new apartment through them, too.

Probably feeling confident that a deal was in the works, the lady agent said: "Why, yes of course. We would be glad to sell your apartment. If you sell and buy through us, you are entitled to *50% discount on the commission.*"

I thought that was rather generous and so I readily acceded to her proposal. In an instant, a calculator materialized out of nowhere, and before she punched the keys she apologized for her weakness in mathematics. "Oh, mathematics is my Waterloo. I was never good with numbers," she declared. I nodded. A few punches and then she said: "Okay, 15% commission on the buying, and 15% commission on the selling, altogether 30%, right? So with a 50% discount on the commission, you will pay only 15% if you buy and sell with us."

What a generous offer indeed, but I had to decline her offer.

On the way home, I had wondered whether she was perhaps the same person who wrote the article in that newspaper on the statistics of Israelis suffering from depression? Your guess is as good as mine.

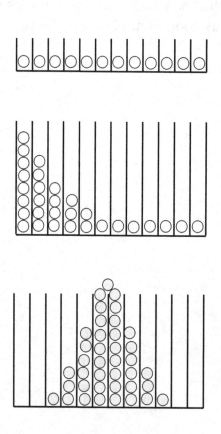

CHAPTER 3

Discover the Uniform Spatial Distribution

In this chapter, we will do some experiments with marbles in cells. In doing these experiments we shall learn to do research in a very rudimentary way, and we shall also have a preliminary glimpse of the nature of the Second Law.

In each of the following subsections we shall perform an experiment with varying numbers of marbles (NM) in different numbers of cells (NC). We shall also follow the "evolution" of the system by recording and plotting a few indices, such as the multiplicities, the probabilities, Shannon's measure of information (SMI), densities of marbles, etc. The experiments are quite simple. I urge you to do the experiments either with real marbles and boxes, or to simulate them on a computer. If you can't do any of these you can at least imagine doing the experiments, and follow the evolution of the system as if they were the results of your own experiments. You can also have a look at some simulations in my site: ariehbennaim.com.

3.1. Initial Warm-Up: Two Marbles in Two Cells

Let us start with a very simple example. We are given two *different* marbles and two cells. We write down all the possible arrangements of these marbles in the two cells (Fig. 3.1).

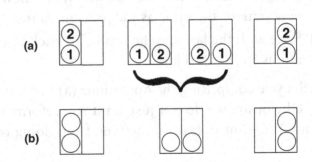

Fig. 3.1 All possible (a) specific arrangements and (b) dim-arrangements of two marbles in two cells.

Altogether, we have four different arrangements. We shall refer to each of these as a *specific* arrangement, specific configuration, specific distribution, specific event, or specific state. It does not matter what term you prefer to use — the important point is to distinguish between these *specific* arrangements and the *dim*-arrangements (or configurations or states).

When we say a *specific* arrangement we mean a *specific* list of which marble is in which cell. There are several ways of describing a specific arrangement. One is shown in Fig. 3.1a, another way is to spell out {*marble 1 in cell one, marble 2 in cell 2*}. When we say a *dim*-arrangement, we do not care which marble is in which cell, but only how many marbles are in each cell. The connection between the two concepts — *specific* and *dim* — is illustrated in Fig. 3.1.

We do the first experiments, with the system described in Fig. 3.1. We choose a starting state or arrangement where all the marbles are in one box. The experiment we do consists of the following operations.

(a) Choose a marble at random. You can do that either by choosing a random integer between 1 and 2 or by tossing a coin. If the result of tossing the coin is a "head," you choose marble (1); if the result is a "tail," you choose marble (2).

(b) Choose a cell at random. Since we only have two cells, you can use the same method of selecting a random integer between 1 and 2, as in (a), but now the result you get is interpreted as the first or second cell.

(c) Place the marble you selected (at random) in step (a) into the cell selected (at random) in step (b).

(d) Now you have a new configuration, or a new *specific* state. This state could be a new state or the same as the previous state.

(e) Start with the last specific result, go back to (a) and repeat the sequence of operations from (a) to (e).

Each cycle comprising the operations (a) to (e) is called a *step*.

Obviously, we do not just want to perform a series of experiments and nothing more: We want to learn something from doing this experiment. Therefore, we shall record some numbers that are characteristic to each of the states, and we shall follow the *evolution* of this characteristic numbers as we proceed with the experiment. We hope that by doing so we may discover some trends, some regularities, or irregularities, and then try to make sense of what we observe, and eventually we might discover a new law of nature. There are many ways of analyzing the results of an experiment. If you are experimenting with a new phenomenon you usually do not know in advance what you might expect to discover, or if you will discover anything at all. Therefore, I shall guide you in the manner of following the "evolution" of the system in time. Here, "time" simply means the sequence of snapshots that you will take of the system as it evolves.

For each of the snapshots, record the following items.

(a) The specific state at the nth step.

(b) The corresponding dim-state (simply obtained by erasing the labels on the marbles).

(c) The number of specific states that compose a dim-state. Denote this number as W. We shall refer to this number as the *multiplicity* of the dim-state. The total number of specific states is four. The total number of dim-states is three, as is shown in Fig. 3.1. The dim-state may be obtained from the specific state simply by disregarding the identity or erasing the labels of the two *marbles*. In the example of Fig. 3.1, the dim-state {1, 1} has two specific states. That is all the counting for this case.

(d) The probability of the dim-state. This is the ratio.[1]

$$\frac{W}{Total\ number\ of\ specific\ states}$$

Call this *PR*, for short.

Suppose that the two marbles, red and blue, were thrown at *random* into the two cells. This means that the red marble may land in the left-hand cell with probability 1/2 and in the right-hand cell with probability 1/2. The same is true for the blue marble. Therefore, the probability of finding a *specific* configuration, say {*blue in the left cell and red in the right cell*}, is 1/4. Another way of arriving at this probability is to start from the total number of possible outcomes (Fig. 3.1a) and assume that each specific outcome has the same probability. Since the sum of the probabilities of all the outcomes must be unity, the probability of each *specific* event is 1/4.

These probabilities may be interpreted in two ways.

The Ensemble Interpretation

Suppose we prepare a million systems, all using the same process of throwing two marbles (one blue, one red) into the two cells at random. An illustration of such an ensemble of systems is shown in Fig. 3.2. What is the fraction of the systems in this

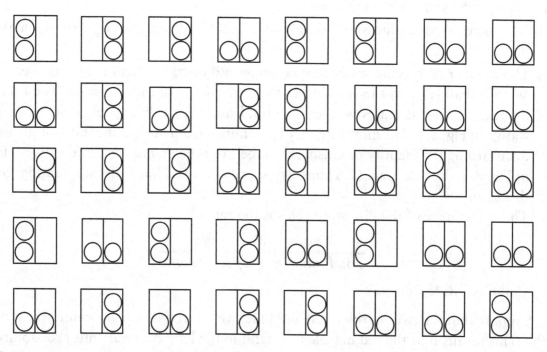

Fig. 3.2   An ensemble of systems, all having the same number of marbles and the same number of cells.

ensemble having a specific state? Clearly, if the ensemble is very large, the fraction of the systems having a specific arrangement is 1/4.

The Frequency Interpretation

Suppose we start with a single system with some specific initial state. We shake the system in such a way that the marbles can cross freely from one cell to another. Every second, we take a snapshot of the state of the system. We produce a series of snapshots as illustrated in Fig. 3.3. Clearly, if we take millions of snapshots and count the *number* of occurrence of each specific state, we shall find that any specific state occurs in one of every four snapshots. Another way of saying the same thing is that the frequency of each specific state is 1/4.

The two interpretations of the probabilities are considered to be equivalent. The system is characterized by giving the *probability distribution* of all the specific outcomes, which we write as (1/4, 1/4, 1/4, 1/4). In this case the probability of *each* specific outcome is 1/4. In all of the following experiments we shall be interested in the dim-state distribution, or the *state* distribution for short. In this example there are three possible state distributions: (1, 0), (1/2, 1/2), and (0, 1).

(e) The average number of questions one needs to ask in the smartest strategy or, for short, the *ANOQONTAITSS*. This is the most important quantity. You only have to imagine that you are playing the 20Q game with this system (see Section 2.3) and count the number of questions. Instead of the long acronym *ANOQON-TAITSS*, we shall refer to this simply as the Shannon measure of information, and denote it *SMI*.

Remember the games in Section 2.3? The *SMI* is simply the number of questions you have to ask (smartly) to find out where a specific marble is, given the dim-state. If you have any doubts about this, go back to Section 2.3. Make sure you know what is meant by *SMI*.

All these numbers are quite easy to calculate for a small number of marbles in a small number of cells. It becomes very tedious for large number of marbles. Therefore, we shall need the help of the computer. The important point to keep in mind is that even when the numbers become very large, there is no new principle involved in the computations. It is the *same type* of calculation, albeit that it requires much more time and effort to carry out.

An example of recorded results for 20 steps is shown in Table 3.1. A simple and efficient way of analyzing the evolution of the system with time — that is, with number of steps — is to draw a graph of the numbers W, SMI, and PR, which characterize the dim-state at each step. An example of such plots is shown in Fig. 3.4. If you have another idea of characterizing the sequence of states by a numerical value assigned to each state, you can plot these, too. However, for our purpose, the three quantities I suggested above are sufficient. In fact, we shall soon see that only two of these are sufficient in studying the evolution of the system with time; the first two — W and PR — are proportional to each other, therefore we shall later drop one of these. The third one — SMI — is conceptually different from the first two. We shall see later that it is related to W and to PR, but at this stage we shall record it as an independent quantity that characterizes the dim-state.

Figure 3.4a shows the values of W calculated for each step. Clearly, there are three different dim-states: $\{2, 0\}$, $\{1, 1\}$, and $\{0, 2\}$. The first and the third each have only one specific state, while the second has two specific states, as shown in Fig. 3.1. In Fig. 3.4a, we see that we have only two values of W — either 1 or 2 — and in this particular run we got, as expected, the value of 1 for about half of the steps and the value of 2 for about half of the steps.

Figure 3.4b shows the probabilities of each of the dim-states. There are only two values of PR: 0.5 for the dim-state $\{1, 1\}$ and 0.25 for either the dim-state $\{2, 0\}$ or $\{0, 2\}$.

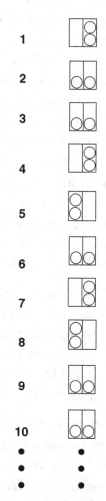

Fig. 3.3 A series of snapshots of a single system being shaken vigorously.

Table 3.1 Recorded Results for 20 Steps of the Game of Two Marbles and Two Cells

	Dim-state	W	PR	SMI
Step 0	{2, 0}	1	0.25	0
Step 1	{2, 0}	1	0.25	0
Step 2	{1, 1}	2	0.5	1
Step 3	{1, 1}	2	0.5	1
Step 4	{0, 2}	1	0.25	0
Step 5	{1, 1}	2	0.5	1
Step 6	{1, 1}	2	0.5	1
Step 7	{2, 0}	1	0.25	0
Step 8	{1, 1}	2	0.5	1
Step 9	{2, 0}	1	0.25	0
Step 10	{2, 0}	1	0.25	0
Step 11	{1, 1}	2	0.5	1
Step 12	{0, 2}	1	0.25	0
Step 13	{1, 1}	2	0.5	1
Step 14	{0, 2}	1	0.25	0
Step 15	{1, 1}	2	0.5	1
Step 16	{1, 1}	2	0.5	1
Step 17	{2, 0}	1	0.25	0
Step 18	{1, 1}	2	0.5	1
Step 19	{0, 2}	1	0.25	0
Step 20	{1, 1}	2	0.5	1

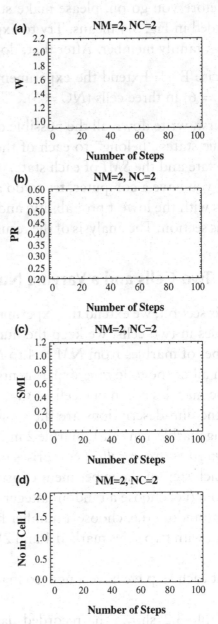

Figure 3.4c shows the values of the *SMI* calculated at each step. Clearly, the state {1, 1} requires one binary question. On the other hand, in both the dim-states {2, 0} and {0, 2} we know the specific state and therefore we do not need to ask any questions. Figure 3.4d shows the number of marbles in cell 1 at each step. Clearly, there are three values: 0, 1, or 2.

Fig. 3.4 The results for the game of two marbles in two cells (2; 2).

Before you go on, please make sure you understand what each of the numbers recorded in Fig. 3.4 means. Try to explain the meaning of these numbers to a friend or to a family member. After that, do the following exercise.

Exercise E3.1: Extend the experiment carried out above to the case of six marbles ($NM = 6$) in three cells ($NC = 3$).

First, write down all the possible dim-states for this system. Calculate how many specific states "belong" to each of the dim-states. Calculate the probability of each dim-state and the *SMI* of each state.

If you have a computer, try to do an experiment by starting with one of the dim-states with the lowest probability, and proceed with the experiment as we have done in this section. The analysis of the results of this experiment is provided in Section 4.1.

3.2. Two Cells and a Varying Number of Marbles

In this section, we extend the experiments we discussed in Section 3.1. Instead of two marbles in two cells, we keep the number of cells fixed ($NC = 2$) and increase the number of marbles from $NM = 4$ to $NM = 100$.

In all of the following experiments, we start with the initial configuration where all the marbles are in one cell — say, cell number 1. Clearly, in this case the specific and the dim-descriptions are the same. Having NM marble in cell 1 is the same as having marble 1 in cell 1, marble 2 in cell 1, and so on. Therefore, the initial dim-state {*all marbles are in cell* 1} comprises only one specific state.

Each step of the experiment consists of three operations: (a) choose a marble at random (i.e. choose a random integer number between 1 and $NM = 4$); (b) choose a cell at random (i.e. choose a number between 1 and 2 — this can be done by tossing a coin, with two sides marked 1 and 2); and (c) place the chosen marble in the chosen cell.

At each step we have a new configuration and we repeat the three steps as stated above.

Table 3.2 shows the recorded data for the case of four marbles in two cells. The first column is the index of the step. The initial configuration is designated

Table 3.2. Recorded Results for 20 Steps of the Game (4;2)

	Dim-state	W	PR	SMI
Step 0	{4, 0}	1	0.0625	0
Step 1	{3, 1}	4	0.25	0.811
Step 2	{2, 2}	6	0.375	1
Step 3	{2, 2}	6	0.375	1
Step 4	{1, 3}	4	0.25	0.811
Step 5	{1, 3}	4	0.25	0.811
Step 6	{1, 3}	4	0.25	0.811
Step 7	{0, 4}	1	0.0625	0
Step 8	{1, 3}	4	0.25	0.811
Step 9	{2, 2}	6	0.375	1
Step 10	{2, 2}	6	0.375	1
Step 11	{1, 3}	4	0.25	0.811
Step 12	{0, 4}	1	0.0625	0
Step 13	{1, 3}	4	0.25	0.811
Step 14	{2, 2}	6	0.375	1
Step 15	{2, 2}	6	0.375	1
Step 16	{3, 1}	4	0.25	0.811
Step 17	{3, 1}	4	0.25	0.811
Step 18	{3, 1}	4	0.25	0.811
Step 19	{2, 2}	6	0.375	1
Step 20	{3, 1}	4	0.25	0.811

as step zero. The second column shows the dim-state in the form {*number of marbles in cell* 1, *number of marbles in cell* 2}. In this notation, the zero configuration is {4, 0}. The third column is the number of specific states that comprise the dim-state. You should be able to check some of the numbers denoted by W in this column. The fourth column is the probability of that dim-state. This probability denoted PR is obtained by dividing W by the total number of specific states.[1] Physically, it means that if we perform many steps — say, 1000 — the fraction of steps at which we observe that dim-state is PR. The fifth column records the Shannon measure of information, denoted SMI. Mathematically, this number is calculated from the distribution of the marbles in the boxes at each step.[2] Intuitively, you already know what the SMI means. If not, go back to Section 2.3 and play some more of the 20Q games.

Because of its central importance, we shall discuss a few examples of how one estimates the SMI. I urge you to examine the following cases carefully. These are quite simple for NM = 4 and NC = 2. It will be much more difficult to calculate these quantities later.

The dim-state {4, 0} consists of only one specific state. Therefore, given the dim-state, we also know the specific state. Hence, no questions needed to be asked. Thus, the value of SMI is 0 (see Table 3.2). The second dim-state is {3, 1}, and the corresponding multiplicity is 4. The probability (PR) is 0.25 and the SMI is about 0.8.

This means that we need, on average, less than one question to find out where a specific marble is. The next case is the dim-state $\{2, 2\}$. Clearly, there are six possible specific arrangements. Therefore, it is more difficult to find out where a specific marble is, given only the *dim-configuration*. The corresponding SMI value for this dim-state is $SMI = 1$; only one question is needed.

Figure 3.5 shows the evolution of the system. Again, I urge you to examine these figures carefully and make sure you understand the meaning of each of the quantities in the figures.

Next, we look at the results with $NM = 8$. Notice again that the graphs of W and PR are very similar. Later, we shall drop one of these since they are easily convertible. Note also that we have started with the configuration $\{8, 0\}$ — that is, all eight marbles in one cell — if you count the number of points at each level you will find that they are proportional to the number of specific states belonging to each dim-state. In this particular case, there are nine different dim-states. Here they are, along with their multiplicities:

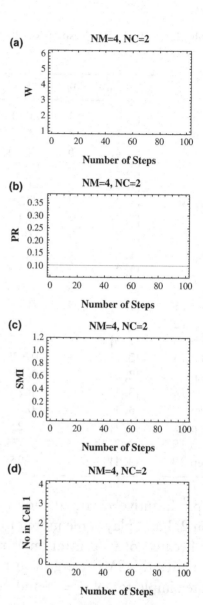

Fig. 3.5 The results for the game of four marbles in two cells $(4; 2)$.

Dim-state	$\{8, 0\}$,	$\{7, 1\}$,	$\{6, 2\}$,	$\{5, 3\}$,	$\{4, 4\}$,	$\{3, 5\}$,	$\{2, 6\}$,	$\{1, 7\}$,	$\{0, 8\}$
The number of specific states	1	8	28	56	70	56	28	8	1

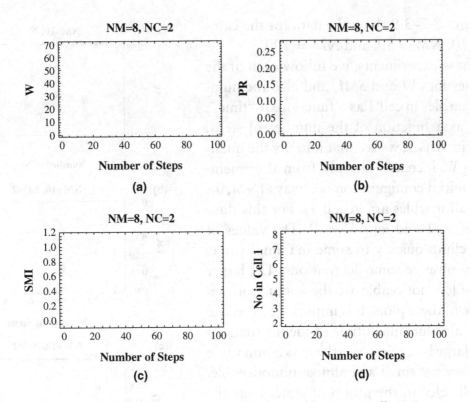

Fig. 3.6 The results for the game of eight marbles in two cells (8; 2).

Count also how many times the system has visited each of the dim-states. Why? Figure 3.6 shows all the data calculated for this case.

In all of the examples so far, we can calculate all the numbers W, PR, and SMI for each dim-state easily. From now on we shall increase the number of marbles, and the calculations will be done by the computer.

Figure 3.7 shows the same graphs for $NM = 10$. Write down all possible dim-states for this case and see if you can also calculate the number W for each of these states.

Note that once we have more than 10 marbles, the numbers W become quite large and the probabilities become very small. Therefore, in the next experiments we shall plot the logarithm to the base 2 of W. We shall also drop the graph of PR. This is simply proportional to W and does not convey any new information.

Figures 3.7–3.9 show the data for the cases $NM = 10$, $NM = 50$, and $NM = 100$.

In these experiments, we follow each of the quantities $\log_2 W$ and *SMI*, and also the number of marbles in cell 1 as a function of "time"; that is, as a function of the number of steps. As NM increases, we see that initially the quantity $\log_2 W$ increases sharply from 0 (remember the initial configuration is always $\{NM, 0\}$; that is, all marbles are in cell 1). For this dim-state, $W = 1$ and $\log_2 W = 0$. The values of $\log_2 W$ climb quickly to some maximal value. Then we observe some fluctuations. The larger NM, the less noticeable are the fluctuations on the scale of these plots. It is important to realize that the absolute magnitude of the fluctuations is quite large but, relative to the maximum value of W, they are small and almost unnoticeable. We shall refer to the group of states near the maximal value of W, where the graph levels off, as the *equilibrium state* of the system.

Note also that the *SMI* behaves similarly; first a sharp increase in the value of *SMI* from 0 towards 1, and then there are some fluctuations, but these become smaller and smaller as NM increases.

When we get to $NM = 100$, all the curves seem initially to be climbing almost monotonically, then leveling off and remaining flat.

Notice also how the number of marbles in cell 1 changes with time. For instance, in Fig. 3.9 we see that initially we have $NM = 100$ in cell 1, then the number drops sharply after 100

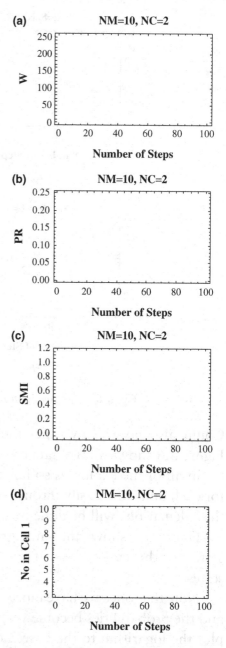

Fig. 3.7 The results for the game of 10 marbles in 2 cells $(10; 2)$.

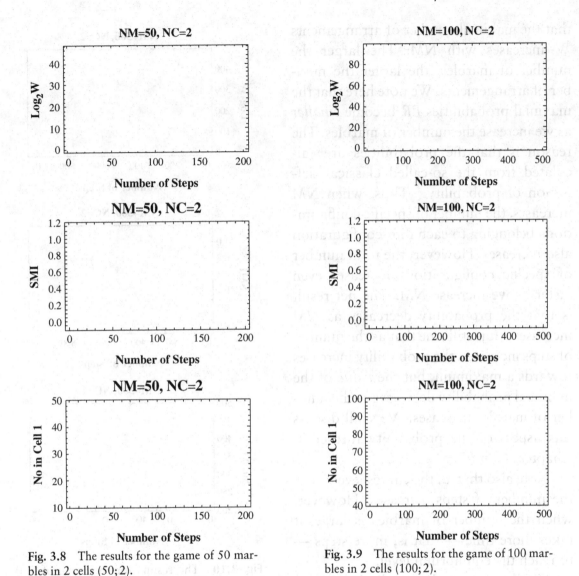

Fig. 3.8 The results for the game of 50 marbles in 2 cells (50; 2).

Fig. 3.9 The results for the game of 100 marbles in 2 cells (100; 2).

steps, and then fluctuates above and below 50, which means about an equal number of marbles in each of the two cells.

Figures 3.10 and 3.11 show the data for $NM = 500$ and $NM = 1000$. As is clear from the figures, the larger the number of marbles, the smoother the curves, and the fluctuation can barely be seen on the scale of these figures. Note carefully that the maximal number of arrangements W increases with NM: The larger the number of marbles, the larger the number of arrangements. We note here that the maximal probabilities PR become *smaller* as we increase the number of marbles. The reason is that the probabilities are calculated from the so-called classical definition of probability.[3] Thus, when NM increases, the number of specific configurations belonging to each dim-configuration also increases. However, the total number of specific configurations increases even faster as we increase NM. The net result is that the probability decreases as NM increases.[3] It is still true that as the number of steps increases, the probability increases towards a maximum, but the *value* of the maximal probability decreases as the number of marbles increases. We shall discuss this aspect of the probability further in Chapters 6 and 7.

Note also that all the curves level off as the number of steps increases. However, when the number of marbles is large, it takes more time — that is, more steps — to reach the equilibrium level.

Finally, we note that the limiting value of the *SMI* for all the cases in Figs. 3.6–3.11

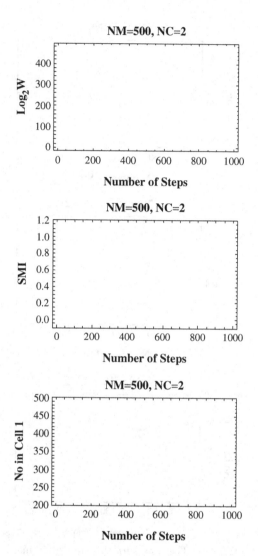

Fig. 3.10 The results for the game of 500 marbles in 2 cells (500; 2).

is the same: unity. The reason is that no matter how many marbles there are, once the average number of marbles in each cell reaches about half of the total number of marbles, we have about equal probability of finding any specific marble in either cell 1 or cell 2, independently of the number of marbles in the system. Again, we

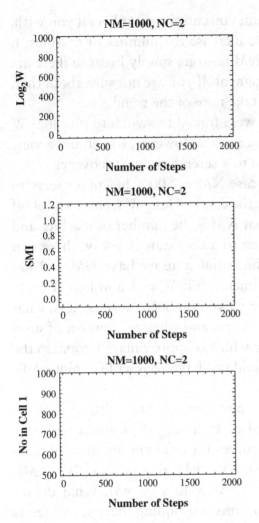

Fig. 3.11 The results for the game of 1000 marbles in 2 cells (1000; 2).

shall discuss this aspect of the *SMI* further in connection with the Second Law in Chapters 6 and 7.

It is time to conclude what we have learned from the experiments in this section. In all the experiments we found that from any initial configuration (not necessarily the one where all marbles were in cell 1), the system will evolve in such a way that the values of *W*, *SMI*, and PR will reach a maximum value. After a large number of steps, the final state will be the one where about half of the marbles will be in each of the two cells. There will be fluctuations about this equilibrium level; the larger the number of marbles, the smaller the relative deviations from this equilibrium level.

Before we leave this section, I want to suggest that you pause and examine carefully the results obtained so far. This has been your first venture into scientific research, and it is important for you to adopt the habit not just of collecting and registering data but also of penetrating into the data. See if you can find some regularities or irregularities, perhaps some common trends. If you do that, you might stumble upon something new that you have not seen before, and you might even discover something no one has ever seen before.

Have you ever heard of the name Darwin? I am almost sure you have. However, had Darwin been satisfied only with collecting and registering an immense amount of data on biological diversity, he would have been one of the most knowledgeable persons about biology, but you would probably never have heard his name.

Let me suggest one way of examining the data; you can choose others if you wish. For instance, look at how W changes when we increase the number of marbles. It is very clear why W becomes very large when NM is large, simply because there are more specific arrangements for each dim-arrangement. If you are not sure about that, try it again on two, three, or four marbles and take note of the trend.

At some point, W grew too big, so that we were forced to switch to plot $\log_2 W$ instead of W itself. This was done for convenience only. However, sometimes a view from a different angle of the same data can lead to a serendipitous discovery!

Look again at the curves of $\log_2 W$ for the case $NM = 500$. The curves seem to level off at about 500. For the case $NM = 1000$, the curve of $\log_2 W$ seems to level off at about 1000. Is this accidental? Remember that NM is the number of marbles and W is the number of specific states corresponding to a dim-state. Clearly, these two numbers are very different. For instance, for the initial state we have $NM = 1000$ (as well as for any step in this particular experiment), but $W = 1$ and $\log_2 W = 0$. Clearly, there is nothing in common between these two numbers. Nevertheless, what we have observed here is that when NM becomes large, and when the number of steps is large, then $\log_2 W$ levels off at a limiting value which is approximately equal to the value of NM. This is something new that we could not have seen had we plotted W itself instead of $\log_2 W$.

Because of the fact that $\log_2 W$ approaches approximately the value of NM for large numbers of steps, I suggest that instead of plotting $\log_2 W$ it might be better to plot $\log_2 W/NM$. This will "normalize" the curves for different numbers of marbles (note that $\log_2 W/NM$ means $\log_2 W$ *divided* by NM; that is, $(\log_2 W)/NM$). Figure 3.12 shows the curves of $\log_2 W/NM$ for two values of NM. What do we see here? The two curves level off at 1. This confirms our finding that $\log_2 W$ tends to NM at very large numbers of steps.

Look further at the curves in Fig. 3.12 and you will see that they are very similar to the *SMI* curves: only "similar," approximately the same, or exactly the same? Well, to examine this question, let us plot both $\log_2 W/NM$ and *SMI* on the same plot. Figure 3.13 shows these two plots on the same graph for $NM = 100$ and $NM = 1000$.

For $NM = 100$, the two plots are quite nearly the same, but for $NM = 1000$ they are almost the same. In fact, even for $NM = 1000$ the two curves are not *exactly* the

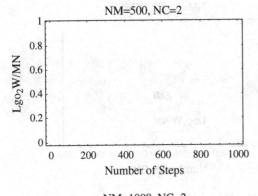

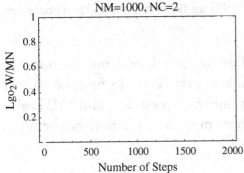

Fig. 3.12 Two plots of the $\log_2 W/NM$ curves.

same. Figure 3.14 shows two amplifications of the plot for $NM = 1000$ at small and large numbers of steps. Now you can see that the curves are almost, but not exactly the same.

Is this accidental? What does that mean? Remember that W and SMI measure two very different things. W is the number of specific arrangements corresponding to a dim-arrangement. SMI is defined as the number of questions we need to ask to find out in which cell a specific marble is. These are truly very different things. Have we discovered something new?

As we saw in Section 2.3, if we have W boxes, the number of questions we need to ask to find out in which box the coin is hidden is $\log_2 W$. We can therefore interpret $\log_2 W$ in terms of the 20Q games. I choose one of the W arrangements (forget

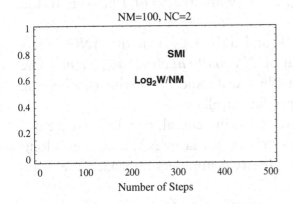

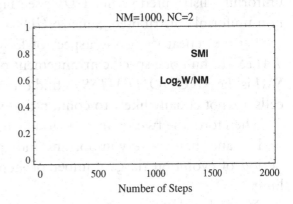

Fig. 3.13 Values of $\log_2 W/NM$ and SMI plotted for the cases $NM = 100$ and $NM = 1000$. Both with $NC = 2$.

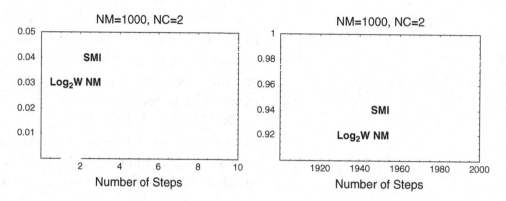

Fig. 3.14   Amplifications of the curves of *SMI* and $\log_2 W/NM$ for the first few initial and final steps. For the case $NM = 1000$, $NC = 2$.

about the coin and the boxes) and you have to find out which one, out of all the W arrangements I have chosen. We saw that if you are smart you can find the answer in $\log_2 W$ questions. So $\log_2 W$ is related to the number of questions, and *SMI* is also related to the number of questions; therefore, there must be a relationship between the two games, right?

Almost right!

Indeed, we cast $\log_2 W$ in the form of the 20Q game, but this game is a very different game from the one associated with the number *SMI*. These are not only two different *games*, but they belong to different classes of games. The first belongs to the uniformly distributed game (*UDG*; see Fig. 2.19), while the second belongs to the non-uniformly distributed game (*NUDG*).

Let me repeat the game aspect of $\log_2 W$ and *SMI*. $\log_2 W$ is the *ANOQON-TAITSS* to find one specific arrangement out of W *equally likely arrangements*. The *SMI* is the *ANOQONTAITSS* to find out in which cell a specific marble is, when the cells are not equally likely to contain that specific marble.

Therefore, the two quantities do not have to be, in general, equal. What we did find — and that is a very important finding — is that for large NM and after a long period of evolution (large number of steps), the two numbers approach the same limit.

Note also that in each of the experiments we have performed in this section, the game has been changed during the evolution of the system. Initially, we had a

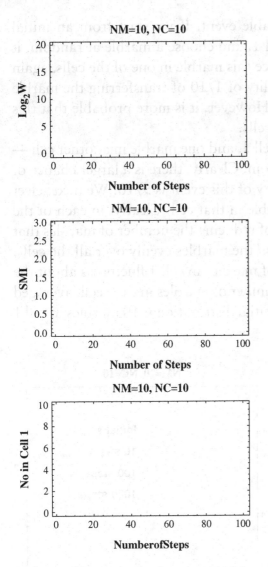

Fig. 3.15 The results for the game of $NM = 10$ and $NC = 10$ (10; 10).

non-uniformly distributed game. Therefore, the initial game belongs to the region denoted *NUDG* in Fig. 2.19. As the system evolves towards the equilibrium game, the point representing the game in Fig. 2.19 moves in the region *NUDG* but eventually reaches a point in the *UDG* region in Fig. 2.19. At equilibrium, the density profile is uniform and therefore the equilibrium game belongs to the region *UDG*. Thus, we can describe the evolution of the games in this and in the next Section as an *excursion* of a point starting in the *NUDG* region in Fig. 2.19, reaching a point in *UDG* region at equilibrium.

We shall see the significance of these findings in relation to entropy and the Second Law in Chapter 7. Until we get there, be patient. We still have to collect a host of additional data.

3.3. Ten Marbles and Varying Number of Cells

Next, we perform a new set of experiments. This time we fix the number of marbles — say, $NM = 10$ — and change the number of cells.

We start with $NM = 10$ and $NC = 10$ (Fig. 3.15). Let us first look at the graph of "numbers in first cell." Recall that we started with all marbles in cell 1. Indeed, the first point at step zero is 10. You can see that this number drops sharply in the first few steps. In principle, it could also climb back to 10, but this is a very improbable event. If we start from an initial configuration where all the marbles are in cell 1, and choose a marble at random, it will necessarily come from cell 1. We then place this marble in one of the cells, again chosen at random. Clearly, there is a probability of 1/10 of transferring the marble to any one of the cells, including cell 1 itself. However, it is more probable that this marble will fall in any one of the *nine* empty cells.

In the next step, we have nine marbles in cell 1, and one marble in another cell — say, cell 7. Again, we choose a marble at random. Clearly, there is a larger chance of selecting the marble from cell 1; the probability of this event is 9/10. We next select one of the cells at random, and place the marble in that cell. Clearly, in each of the first few steps there will be larger probability of reducing the number of marbles that were initially concentrated in cell 1, and *spread* the marbles evenly over all the cells.

After many steps we see that the number of marbles in cell 1 fluctuates about the value of one. Figure 3.16 shows the average number of marbles in each cell, averaged over 10 steps, 100 steps, and 1000 steps. The initial distribution is 10 marbles in cell 1 and 0 in all others. After 10 steps, we see a reduction of the average number of marbles in cell 1, and eventually after 100 and 1000 steps we see a uniform distribution or maximal spread of the marbles over the 10 cells. We shall refer to the curves in Fig. 3.16 as the density profile of the system. The density profile obtained after many steps (here 1000) may be referred to as the equilibrium density profile of the system.

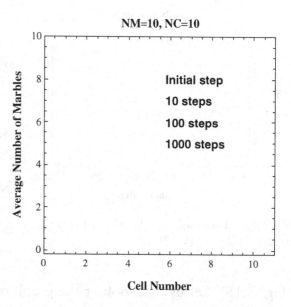

Fig. 3.16 Density profile of the system with *NM* = 10 and *NC* = 100 (10; 100).

The curves of *W*, *PR*, and *SMI*, all look the same, but they signify different aspects of the evolution of the system. The initial value of *W* is clearly 1 (hence, $\log_2 W = \log_2 1 = 0$). There is only *one* specific

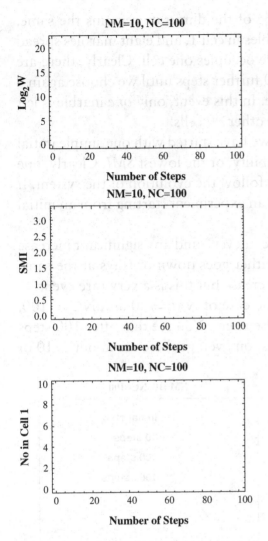

Fig. 3.17 The results for the game of $NM = 100$ and $NC = 10$ (10; 100).

arrangement that belongs to this dim-state. The probability of the initial specific state relative to all other possible specific states is very small, therefore PR is very small and $\log_2 PR$ is large and negative. The initial value of SMI is 0. Why? Simply because in the initial state, we *know* where all the marbles are, therefore we have to ask *zero* questions in the 20Q game. As the number of steps increases, all the graphs initially climb upwards, and after about 30–40 steps there is a kind of "leveling off," but with noticeable fluctuations above and below some average value.

If you have any doubts as to the meaning of these curves and why they have this particular form, try to work out a simpler case — say five marbles in five cells — or do the experiments on your computer, following the evolution of the system and rationalizing the changes at each step.

Next, we turn to the case of 10 marbles in 100 cells. Again, I urge you to examine carefully all the curves in Fig. 3.17 and compare these with Fig. 3.15.

First, look at the curve of "number of marbles in the first cell." As you can see, we started with all marbles in cell 1, and after a few steps this number declines sharply from 10 to 0. But note that at the level of 2, there are quite a few points before it goes down further. This is because when there are only *two* marbles left in cell 1, the next marble to be chosen at random is more likely to come from the eight marbles that are already spread over the 99 cells. In the next few steps, it is likely that, though the configuration will continue to change, the type of the dim-state remains the same. These dim-states are characterized by *two* marbles in cell 1, and eight marbles spread over the 99 cells in such a way that one marble occupies one cell. Clearly, there are many of these dim-states, and it takes about 10 further steps until we choose again a marble from cell 1 and place it somewhere else. In this event, only *one* marble is left in cell 1 and the rest are evenly spread over all other 99 cells.

In all of the experiments carried out so far, we have started with one simple initial configuration — the one with the lowest probability, or the lowest *SMI*. Clearly, one can start from any arbitrary configuration and follow the evolution of the system. If you have any doubts about this, try to simulate an experiment starting from an initial configuration of your own choice.

Note also that in contrast with Fig. 3.15, we "never" find any significant increase in the number of marbles in cell 1; the curve either goes down or stays at the same level. Of course, there is some probability of increase but this is a very rare event.

Figure 3.18 shows the density profile for the case of $NM = 10$ and $NC = 100$. The initial step is 10 marbles in cell 1, and 0 elsewhere. You see that after 100 steps the marbles are evenly distributed over the cells; on average there are about 1/10 of marbles in each cell. You notice small fluctuations about this average value, but they are not significant.

Figure 3.19 shows the data for the case of $NM = 10$ and $NC = 1000$. Look first at the number of marbles in cell 1. The value steeply declines from the initial value of 10 to 0. Figure 3.20 shows the density profile for this case.

All the curves of $\log_2 W$, $\log_2 PR$ and *SMI* are similar to each other. Here, all the curves are "smoother"; that is, fluctuations become much smaller. When the number of either the marbles or the cells becomes very large, you will not be able to *observe*

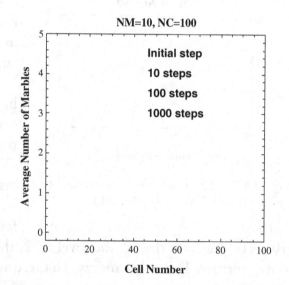

Fig. 3.18 Density profile for the game of $NM = 10$ and $NC = 100$ (10; 100).

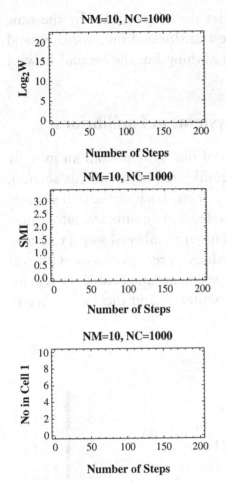

Fig. 3.19 The results for the game of $NM = 10$ and $NC = 1000$ $(10; 1000)$.

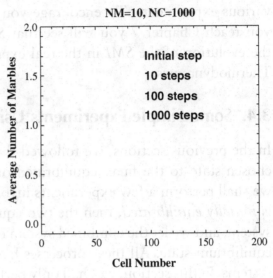

Fig. 3.20 Density profile for $NM = 10$ and $NC = 1000$ $(10; 1000)$.

any fluctuations; the curves will be monotonic and once they level off they will stay there "forever."

"Forever" here means forever in your lifetime, or in billions of years. Small fluctuations will always occur, but they are relatively small and almost impossible to observe or to measure. On the other hand, large fluctuations are in principle observable and measurable, but they are so improbable, perhaps occurring once in a billion, billion, billion years, that in practice they are never observable.

In the next section, we shall do a few other "experiments" which are the analogs of some real experiments, such as expansion of gas, mixing of two gases, etc.

Before doing that I advise you to pause and reflect on what you have learned so far. Check again your understanding of all the quantities that were recorded and plotted in the figures. In particular, check your understanding of the meaning of *SMI* and take note of what you have learned about the evolution of this quantity in the various experiments. To encourage you to do so, let me tell you that by the time you reach Chapter 7 you will see that *SMI* will be transformed into *entropy*, and the evolution of the *SMI* in the real experiments is nothing but the Second Law of Thermodynamics.

3.4. Some Simple Experiments Using Two Systems at Equilibrium

In the previous sections, we followed the evolution of *one* system from an initially chosen state to the final, equilibrated or nearly equilibrated state. In this section, we shall perform a few experiments involving two systems. Each of the two systems is *initially equilibrated*, then the two equilibrated systems are combined into a new system, and we let the combined system evolve in time (or number of steps) to a new equilibrium state. All these processes have their analogs in real processes with real systems. In this section, we shall only perform the experiments and record the results. The interpretation of these results will be left to Chapter 6, and their relevance to real processes will be further discussed in Chapter 7.

3.4.1. *A Process of Expansion*

Figure 3.21 shows an experiment analogous to an expansion process. Initially, we have 50 marbles equilibrated in 5 cells, and another system of 5 empty cells (Fig. 3.21a). The two systems are brought into "contact" (Fig. 3.21b). We start the process as before, and follow the evolution of the combined system of 50 marbles in 10 cells. The evolution of the parameters *SMI* and $\log_2 W/NM$ is shown in Fig. 3.22. Note carefully that the initial value of *SMI* is about 2.32 (not 0 as in the previous experiments). The final value is 3.32 (these numbers correspond to $\log_2 5 \approx 2.32$ and $\log_2 10 \approx 3.32$, respectively). The change in *SMI* is *one unit*. Explain this result in terms of the 20Q game.

Figure 3.23 shows the average number of marbles in each of the cells. Initially, the curve is a step-function;

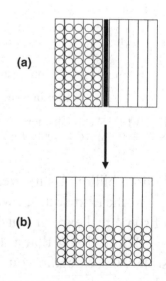

Fig. 3.21 The process of expansion of 50 marbles, initially in 5 cells to 10 cells.

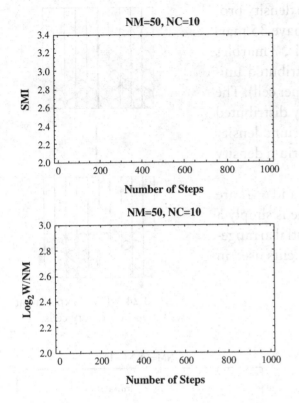

Fig. 3.22   The evolution of *SMI* and $\log_2 W/NM$ for the expansion process.

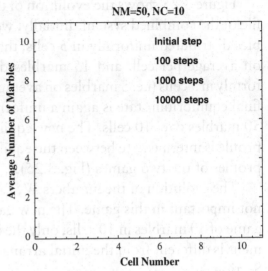

Fig. 3.23   The evolution of the density profile for the expansion process.

that is, the value is 10 for cells 1–5, and 0 elsewhere. As the number of steps increases, we reach a new equilibrium state for the *combined* system of 50 marbles in 10 cells. After about 10,000 steps, we observe an almost uniform distribution of marbles, on average about 5 marbles in each cell. You should realize that this experiment is equivalent to an experiment where *NM* is fixed, and we increase the number of cells. The only difference between this and the experiments in Section 3.2 is the choice of the initial state. In Section 3.2, we always started with all marbles in cell 1. Here, on the other hand, the initial state is an equilibrated system of 50 marbles in 5 cells.

3.4.2. *Combination of Two Non-Empty Games*

We next perform an experiment with two non-empty games — say, 35 marbles in 5 cells, and 15 marbles in another 5 cells (Fig. 3.24). We first bring each of the systems to equilibrium. After combining the system, they are brought to a new equilibrium state. The new system consists of *NM* = 50 and *NC* = 10.

Figure 3.25 shows the evolution of the density profile of the combined system. Initially, we have 35 marbles distributed uniformly in 5 cells; that is, 7 marbles on average per cell, and 15 marbles distributed uniformly in 5 cells (i.e. 3 marbles on average per cell). The final equilibrium state is again a uniformly distributed 50 marbles over 10 cells. The new equilibrium density profile is intermediate between the equilibrium density profiles of the two games (Fig. 3.25).

The evolution of the numbers *W*, *PR*, and *SMI* are not important in this game. The new game is simply a game of 50 marbles in 10 cells, only the initial arrangement is different from the initial arrangements used in Section 3.2.

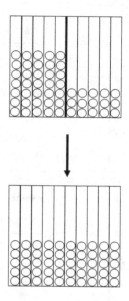

Fig. 3.24 The experiment with two non-empty cells.

3.4.3. *Pure Mixing of Two Types of Marbles*

The following process is the analog of an experiment known as a *reversible mixing* of two gases (Fig. 3.26).

In this process, we combine two kinds of marbles — say, 32 blue marbles in 16 cells and another 32 yellow marbles in 16 cells — to form one system of 32 blue marbles and 32 yellow marbles in 16 cells. At equilibrium we have on average two blue marbles per cell in one system, and two yellow marbles per cell in the second system. Note carefully that in this process the final number of cells is 16, not 32. Compare this with the process in Section 3.4.5 below.

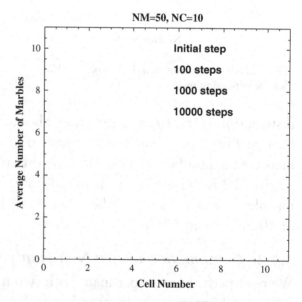

Fig. 3.25 The evolution of the density profile for the two non-empty cells.

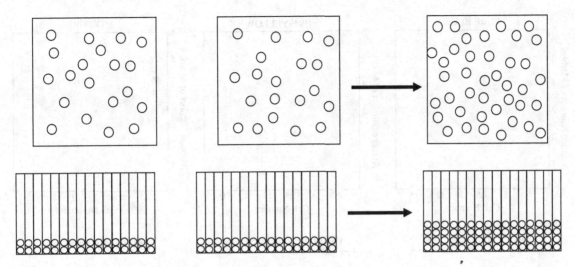

Fig. 3.26 Illustration of the reversible mixing of two real gases, and the analog process with marbles in cells.

Figure 3.27 shows the equilibrium density profile of the pure systems and the combined system. Clearly, we observe mixing of the two colored marbles. At equilibrium, we find on average about two blue marbles and two yellow marbles per cell in the combined system.

Figure 3.27 also shows the evolution of the *SMI* in this process. First, we show the evolution of the blue marbles and the yellow marbles separately. The plot on the right-hand side shows the evolution of the mixture. Note that the *SMI* of the mixture is the *sum* of the *SMI* of the separated equilibrated system, but the value of *SMI* does not change after combining the two systems and reaching the new equilibrium state.

As I have indicated in the beginning of this subsection, this process is the analog of a reversible mixing of two gases. You do not need to know what "reversible mixing" means. At this stage you are required only to carry out the experiment, or imagine doing the experiment, and record the results. Note carefully that the initial value of the *SMI* of the *combined* system (i.e. the left-hand side of Fig. 3.26) is about $4+4 = 8$. In the final mixture (right-hand side of Fig. 3.26), the value of *SMI* is also about 8.

Can you give a rationale for the value of $SMI = 4$? (Note that $\log_2 16 = 4$.) Can you explain why the value of the *SMI* of the combined system does not change in this process?

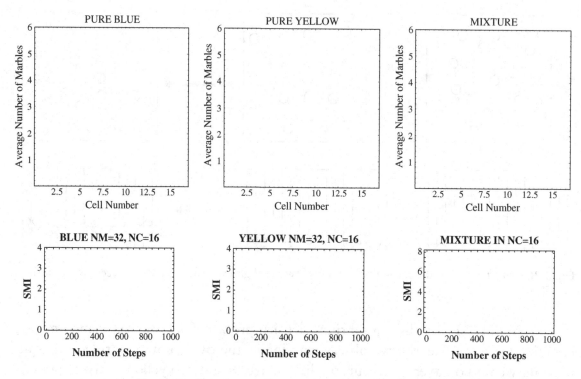

Fig. 3.27   Density profiles for the pure mixing process.

3.4.4. *Pure Assimilation Process*

This process is similar to the process of pure mixing as shown in Fig. 3.26, but now all the marbles are of the *same kind*. We start with 32 marbles in 16 cells and another 32 marbles in 16 cells. We combine the two systems into a single system consisting of 64 marbles in 16 cells. Figure 3.28 shows the process and the density profile of the initial two systems and of the final combined system. At equilibrium, we find on average four marbles per cell, compared with two marbles per cell in each of the initial systems. Also, we find that in this experiment, the *SMI* of the final combined system is smaller relative to the initial combined state. Can you explain why? We shall discuss this experiment further in Chapters 6 and 7.

3.4.5. *Mixing and Expansion*

This process is similar to the process of expansion discussed in Subsection 3.3.1. We start with 32 blue marbles in 16 cells and 32 yellow marbles in 16 cells, and combine

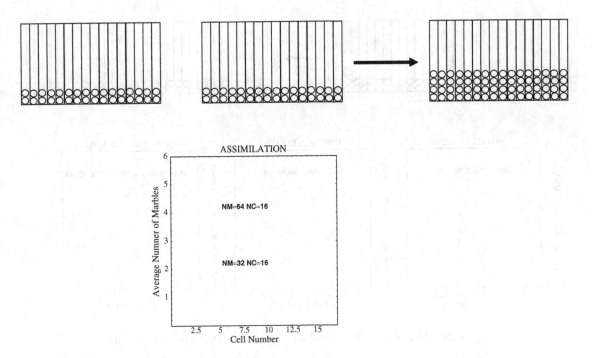

Fig. 3.28   The process of pure assimilation and the density profiles for the process.

the two systems to obtain one system of 32 blue marbles and 32 yellow marbles in 32 cells. In Fig. 3.29 we also show the density profile of this system, before and after the mixing process.

It should be noted that the average number of marbles (either blue or yellow) per cell decreases from two per cell to about one per cell. This process is essentially an expansion of the blue marbles from 16 cells to 32 cells, and the same for the yellow marbles.

This process is of importance in the study of thermodynamics. We shall discuss this process further in Chapters 6 and 7.

3.4.6. *Assimilation and Expansion*

This process is similar to the one described in Subsection 3.3.5, but instead of two *different* kinds of marbles, we have only one kind. We start with two systems, each with 32 marbles in 16 cells, and combine them into one system of 64 marbles in 32 cells. Note the difference between this process and the processes described in Sections 3.4.4 and 3.4.5.

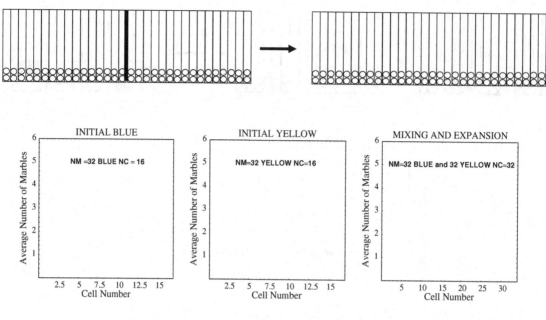

Fig. 3.29   Density profiles for the process of mixing and expansion.

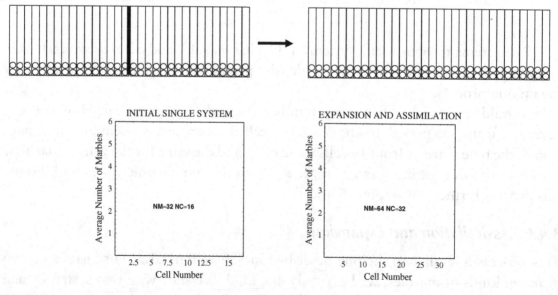

Fig. 3.30   Density profiles for the process of assimilation and expansion. **(a)** The single system before the process. **(b)** The combined systems after the process.

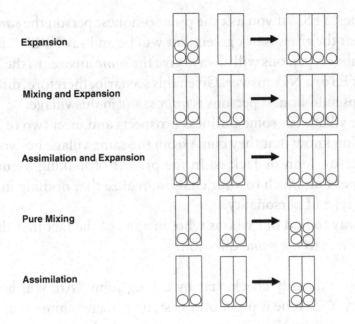

Fig. 3.31 The process for exercise E3.2.

Figure 3.30 shows the process and the density profile of this process. Clearly, there is not much that changes in this process. Both in the initial and in the final states, we have at equilibrium an average of about two marbles per cell. We shall discuss this process further in Chapters 6 and 7.

Exercise E3.2: Calculate the changes in *SMI* and *W* for the five processes shown in Fig. 3.31.

Snack: How to Tell the Difference Between an Honest and a Pseudo-Honest Person (adapted from Mero (1990))

Not far from the Amazon region lie lush and verdant hills and valleys. At the foot of one of these valleys lies a village inhabited by only two types of people: the honest and the pseudo-honest. The honest people always see the world as it actually is, and they never lie. The pseudo-honest people, on the other hand, see the exact opposite of what is reality and they never tell the truth. Asking an honest person "Is the grass green?"

the answer will be "YES." If you ask the pseudo-honest person the *same* question, he or she will *see* that the grass is not green, but will lie and say "YES." Both the honest and the pseudo-honest persons will always give the *same* answer to the *same* questions which require a YES or a NO answer. Given this scenario, therefore, differentiating the honest from the pseudo-honest persons is *impossible* in this village.

You go to the village for some business prospects and meet two of its inhabitants, John and Jack. You know that they come from the same village but you do not know which type of person John or Jack is. In the pretext of making friends, you talk to them, probe deeper but, much to your chagrin, realize that nothing in their behavior gives away their type of personality.

Is there any way to find out who is who, in spite of the fact that they will always give the *same* answer to the *same* question?

Answer: Yes, you can easily check that by asking John "Are you honest?" If he is honest, he will say YES; if he is pseudo-honest, he perceives himself as honest, but he will lie and answer with a NO. The same applies to Jack, who will answer YES if he is honest and NO if he is pseudo-honest.

Therefore, by asking anyone of them "Are you honest?" you will know who is who.

That however, seems to contradict the conclusion reached before — that the two persons of different types will always give the *same* answer to the *same* question. So, how come they gave different answers to the *same* question?

The answer is simple. The two questions are not the same. The *words* are the same, "Are you honest?" but the "you" in the two questions is different. In other words, your question to John is "Are you, John, honest?" This is different from the question addressed to Jack: "Are you, Jack, honest?"

Another solution to the apparent paradox is to change the conclusion that each person of the village gives the *same* answer to the *same* question. Instead, we know from the description of the behavior of the persons of the village that each person answers with the *truth* to each question.

For instance, if you ask "Is five equal to five?" the honest answer will be YES. The pseudo-honest will also answer YES. If you ask "Is five smaller than two?" clearly the honest will say NO, and the pseudo-honest will say NO, too.

From this conclusion, it follows straightforwardly that if you ask the honest person "Are you honest?" he will speak the truth — YES. If you ask the pseudo-honest "Are you honest?" obviously he will also speak the truth and say NO.

Snack: Stray and Pay

Every evening, the same 10 women of Baghdad would go to the central park, sit (in a circular sitting position) and gather around a beautiful fountain while their husbands stood behind them, to enjoy a whiff of crisp spring air, do some mundane tasks such as knitting or crocheting, or simply to relax. Although these women see each other every night they do not talk to each other. While all the women had a vantage view of all the other women and those women's husbands, they did not have the luxury of seeing their own husbands, who stood behind them. As the story unfolds, the fact that the wives could not see their own husbands led some of the husbands to take advantage of the situation.

One evening late in April, Judge Hussein goes to the park, stands beside the fountain and, as if surveying the area, looks around intently. He looks at *all* 10 women and their husbands who stand behind them. Unlike the women who could see everyone except for their own husbands, Judge Hussein saw and witnessed everything. In fact, he saw *two* of the husbands flirting with other women (not in the circle), literally behind the backs of their wives.

Upon seeing this, Judge Hussein got so furious that he decreed a new law on the spot: If a woman catches her husband being unfaithful to her, she must kill him before day breaks the next morning. The women were also warned neither to talk to each other nor discuss what they saw. However, the twist was that he only told the women about the new law, and not the husbands.

Upon hearing this, each of the 10 women looked at all the other women in the circle. Some of the women saw one husband flirting, while some saw two husbands flirting, but unfortunately no women could see the "activities" of their own husbands.

The following evening, Judge Hussein goes to the park and, seeing once again the "activities" of the same "unfaithful" husbands, reminds the wives about the new law and then leaves.

On each of the following evenings, the same two men were flirting with other women. The following morning, no one was killed for the simple reason that the wives could not see what went on behind their backs, and so it went on day after day — no husband was caught with his activities.

After a month had passed, Judge Hussein was enraged as it was a blatant display of the women's defiance of his law. On June 1, he went to the park and again talked to the women only. "Don't you see that there is *at least* one husband who is unfaithful?" he asked rhetorically and, without waiting for an answer, left the park.

The next morning, June 2, as usual nothing had happened. But on the morning of June 3, the two unfaithful husbands were killed by their wives.

Why did the two wives kill their unfaithful husbands only on the second morning? Why did they have to wait for over a month after the law was declared, when each of the women knew exactly what the judge said; namely, that *"at least one husband was unfaithful"*?

Answer: The story is a typical story, here more dramatized, concerning the difference between private knowledge and common (or mutual) knowledge. Indeed, every evening all the women with no exceptions saw and knew that *"there was at least one unfaithful husband."* But they could not *use* this information to find out what their own husbands were up to.

When Judge Hussein came and said *"there is at least one unfaithful husband,"* they did not receive any *new* information aside from what they already knew.

Indeed, no new information was *contained* in the statement that the judge made, and yet some new information could be *inferred* from his statement. The *same* information that every single woman knew was now made public or *common knowledge.* Before the judge made the statement, every woman knew that *"there was at least one unfaithful husband,"* but they did not know that the other women knew that. After the judge made his statement, the women not only *knew* that *"there is at least one unfaithful husband,"* but also knew that each of the *other* women knew that *"there was at least one unfaithful husband,"* and that each of them knew that each woman knew. That additional information could be made useful.

After the first night, as usual nothing happens, although each one saw there was at least one unfaithful husband. However, the next evening, a woman whose husband was unfaithful (but still does not know that) — let us call her Unlaquila (wife 1) — sees that one husband was flirting behind the back of Unfortuna (wife 2) as usual. But now Unlaquila also knows that Unfortuna also heard what the judge said two nights ago. Now, Unlaquila's reasoning was as follows: If during the first night Unfortuna *did not see* any husband flirting, she must have concluded that the "at-least-the-one-husband" was her own husband. In that case she must have killed her husband the first morning after Judge Hussein's visit on June 1. The fact that nothing happened on the first morning meant that Unfortuna *did see one* unfaithful husband. Therefore, nothing happened on the first morning. However, if Unfortuna *saw one* unfaithful husband — Unlaquila continued to reason — it must be my own husband, because there are no other unfaithful men except her own husband, who she could not see. Therefore, Unlaquila concluded that her husband must be the unfaithful one who Unfortuna saw the next evening.

Clearly, the same reasoning goes into the head of Unfortuna, and therefore both Unlaquila and Unfortuna concluded that their husbands were unfaithful and, on the second morning, killed their husbands.

One can prove by mathematical induction that for any number of unfaithful husbands — say, five husbands — will all be killed five mornings after the judge declares that "there are at least four unfaithful husbands."

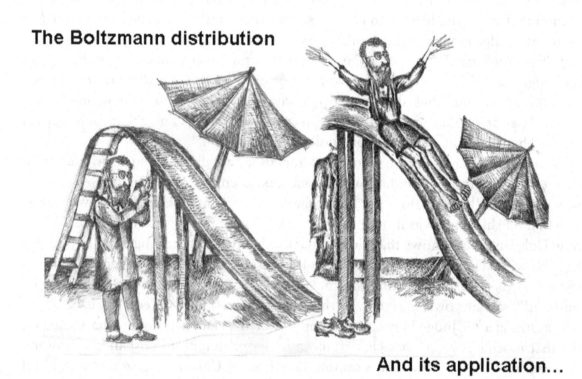

CHAPTER 4

Discover the Boltzmann Distribution

As I told you at the end of Section 3.2, the experiments we carried out in Chapter 3 are sufficient for catching a glimpse of the Second Law. In this and in the next chapter we shall do more experiments under new constraints. These experiments will certainly broaden your horizons. However, you should be informed that they are not essential to the understanding of entropy and the Second Law of Thermodynamics.

If you are interested only in a basic understanding of what entropy is and why it always changes in one direction in spontaneous processes, you can skip this and the next chapter, and instead read only the relevant sections in Chapters 6 and 7. However, I urge you to read these two chapters. Although their reading requires some effort, you will learn two of the most important probability distributions in physics in general, and in thermodynamics in particular. In addition, the last three sections of this chapter discuss some fundamental processes, the analogues of which in real life were the foundations on which the Second Law of Thermodynamics was established.

In the previous chapter, we followed the evolution of games consisting of NM marbles in NC cells. We found that after a large number of steps, all the games evolved and reached a state of equilibrium. This state of equilibrium is not a single state but a group of states that are not far from each other. We have also seen that at equilibrium the distribution of marbles over the cells is uniform; that is, maximal spreading of the marbles over all the cells. The larger the number of cells, the larger the spreading of the marbles. Once reached, the uniformity of the distribution is maintained.

In this chapter, we shall do the same kind of experiments with almost the same kind of game, but with strings attached. What we shall discover is that, as before, the system will evolve towards the state of maximal SMI, which is also the state having the maximal number of specific arrangements. However, unlike the processes *in Chapter 3, the equilibrium distribution will not be uniform. We shall discover the* Boltzmann distribution, *which is of central importance in statistical thermodynamics. The strings attached will also spawn a few new experiments involving exchange of strings. The Boltzmann distribution is a special result that follows from a more general theorem proved by Shannon.*

The strings attached *in our experiment are literally strings attached to the marbles. Do not worry about the significance of these attached strings. In Chapter 7, we shall discuss the motivation for inventing this game, and what the strings stand for. For the moment, treat this game as if it was a little harder than the games we played in Chapter 3, and therefore a more challenging one.*

Imagine that each marble in the system is attached to a string. All the strings are connected to the left side wall of the first cell (Fig. 4.1). The length of the string attached to the marbles in cell 1will be 0. A marble with string length 0 is said to be in "ground zero." A marble in cell 2 will have a string of length 1. We can refer to this length as level 1; *for the marble in the 10th cell, the level is 9, and so forth.*

We will be doing the same experiment as before; that is, starting with some initial configuration, we choose a marble at random and move it to a randomly chosen cell. But now we require the following condition: The total length of the strings must be conserved. This means that when a marble is moved from one cell to another — say, from level 2 to level 3 — its string length changes from 2 to 3, therefore in order to conserve the total string length, another marble must be moved to a lower level. An example is shown in Fig. 4.2.

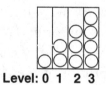

Level: 0 1 2 3

Configuration: {1,2,3,4}
Number of Marbles: 10
Total string length:1x0+2x1+3x2+4x3=20

One unit of length

Two units of length

Fig. 4.1 A system of 10 marbles in 4 cells. Each marble is anchored by a string connected to the extreme left wall. The total string length is 20. The strings are shown only for a few marbles.

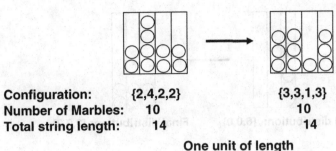

Configuration: {2,4,2,2}   {3,3,1,3}
Number of Marbles: 10   10
Total string length: 14   14

One unit of length

Total string length 14 units

Fig. 4.2 A possible process from one configuration to another, conserving the total string length, here of 14 units.

The protocol for the experiments is the same as in Chapter 3, except that we accept only configurations that conserve the total length of the strings. A few simulations are shown in the site: ariehbennaim.com.

4.1. Six Marbles in Three Cells

Remember exercise E3.1 from Section 3.1? Now we shall discuss that experiment in detail, and modify it for our requirements in this chapter. First, we will do the experiments with no strings attached. We have six marbles ($NM = 6$) in three cells ($NC = 3$), as shown in Fig. 4.3. As you found in that exercise, altogether, there are 729 specific configurations but only 28 dim-configurations. These are grouped into different types — a, b, c, d, e, f, and g — in Fig. 4.3. Each class of dim-configurations has a specific multiplicity; that is, the number of *specific* arrangements that belong to the dim-arrangement. We also calculate the total number of specific configurations for this example, which is 729. In Chapter 3, we were interested in the evolution of this system, starting with any one of these configurations — say, the one of type a on the left-hand side of Fig. 4.3. For this example, we show in Fig. 4.4 the values of SMI, the W, and the evolution of the average number of marbles in each cell. Note specifically that the equilibrium density profile is uniform, as were all the density

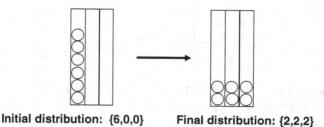

Initial distribution: {6,0,0}   **Final distribution: {2,2,2}**

Dim distributions	Multiplicity	Total
a. {6, 0, 0} {0, 6, 0} {0, 0, 6}	1	1 × 1 = 3
b. {5, 1, 0} {5, 0, 1} {0, 5, 1} {1, 5, 0} {1, 0, 5} {0, 1, 5}	6	6 × 6 = 36
c. {4, 2, 0} {4, 0, 2} {0, 4, 2} {2, 4, 0} {2, 0, 4} {0, 2, 4}	15	6 × 15 = 90
d. {3, 2, 1} {3, 1, 2} {1, 3, 2} {2, 3, 1} {2, 1, 3} {1, 2, 3}	60	6 × 60 = 360
e. {4, 1, 1} {1, 4, 1} {1, 1, 4}	30	3 × 30 = 90
f. {3, 3, 0} {3, 0, 3} {0, 3, 3}	20	3 × 20 = 60
g. {2, 2, 2}	90	90
Total number of specific configurations:		**729**

Fig. 4.3   All possible dim-configurations of a system of six marbles in three cells (6;3). Each type of dim-configuration has a typical multiplicity. The total number of specific configurations is 729. The total number of dim-configurations is 28. The initial and the final configurations are shown on top of the figure.

profile of the experiment in Chapter 3. Note also that there are seven different values of W, as can be seen in Figs. 4.3 and 4.4.

We now attach the strings to each of the marbles in Fig. 4.3. Clearly, different configurations have different total lengths of the strings. From all the dim-configurations listed in Fig. 4.3, we select all the configurations that have a fixed total string length. For instance, for *total string length* $TL = 4$, there are only three configurations. These are shown in Fig. 4.5b. Check that these are only three possible dim-configurations fulfilling the condition $TL = 4$.

Before we do the experiment, make sure you understand the meaning of the values of W and *SMI* for each of these configurations. W is the number of specific configurations belonging to a dim-configuration. For example, the first and third dim-configurations in Fig. 4.5b each have 15 specific configurations. The second dim-configuration has 60 specific configurations. Note also that the latter also has the largest value of *SMI*. Why? To find out, simply label the marbles with numbers from 1 to 6 (or with letters or colors, or anything you wish), and count how many specific configurations belong to this dim-configuration. You should be able to do this exercise and make sure you understand the difference between the *specific* and the *dim* configurations. In the *specific* configurations, the marbles are labeled. The *dim*-configuration is obtained by erasing the labels on the marbles (numbers, colors, or whatever). The multiplicity of a dim-configuration is the number of *specific* configurations that produce the same dim-configuration when the labels are erased.

Now we start the experiment. Since we have only three possible dimconfigurations, the game is comparatively easier than in the previous chapter in the sense that we have fewer configurations, but it is also more difficult because at each step we have to make sure we keep the total length of the string constant (here $TL = 4$). Note that if we move only one marble, the total length cannot be conserved. Therefore, we move two marbles at each step. We check: If the total length is 4, we accept the new configuration; if not, the new configuration is rejected. You could

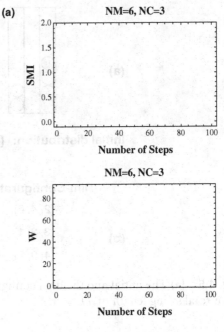

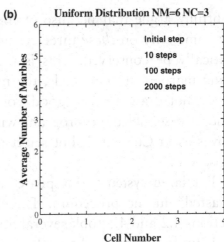

Fig. 4.4 (a) The evolution of the system in terms of *SMI*, *W* and (b) the average number of marbles in each cell, for different numbers of steps.

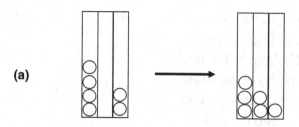

(a)

Initial distribution: {4,0,2}   Final distribution: {3,2,1}

Dim-Configuration	Multiplicity (W)	SMI
{2, 4, 0}	15	0.918296
(b) {3, 2, 1}	60	1.45915
{4, 0, 2}	15	0.918296

Fig. 4.5 (a) The initial and the final configurations for the game (6;3). (b) All possible configurations with total string length of $TL = 4$.

rightfully argue that because of the small number of configurations, we can play the game *only* on *these* three configurations, and the total string length will automatically be conserved. This is true for this particular example. However, with many marbles in many cells, the number of configurations becomes so large that we cannot draw all the "good" ones and play only with these. Therefore, in the general case, start the program with a given dim-configuration, make a random moves as in Chapter 3, but select only those moves for which the string length is conserved.

For large systems, this procedure requires a large number of moves that are "wasted" in the procedure. Therefore, some of the experiments carried out in Sections 4.2 and 4.3 took several hours on a PC.

Figure 4.6a shows the evolution of this system. Note that W oscillates between two values, 15 and 60. These are the multiplicities of the configurations shown in Fig. 4.5b. Similarly, the values of *SMI* oscillate between 0.92 and 1.46. Make sure you understand what these numbers are; if not, play the 20Q game on each of these configurations. Label the marbles by numbers, I will think of a number and you have to find out which cell contains the marble bearing the number I am thinking of. You should also be able to estimate the average number of questions you need to ask in each case. Why is the second configuration in Fig. 4.5b more difficult to play than either the first or the third?

Finally, look at the profile of the average number of marbles in each cell (Fig. 4.6b) and compare this with Fig. 4.4b. We shall have more to say on this particular profile in the next sections. In this figure, we also added three black dots, which are the theoretical values of the limiting distributions.[1]

4.2. Nine Marbles in Varying Number of Cells

We now perform experiments with fixed number of marbles — here, 9 ($NM = 9$) — and three different numbers of cells ($NC = 4, 6, 8$). The specific numbers of marbles and cells were chosen after a long and tedious search for a small system that manifests the new behavior.

We chose a total length of $TL = 9$. The initial configuration corresponding to this length is shown on the left-hand side of Fig. 4.7. In the initial configuration, nine marbles are placed in the second cell, each contributing a unit of length, and therefore the total length is

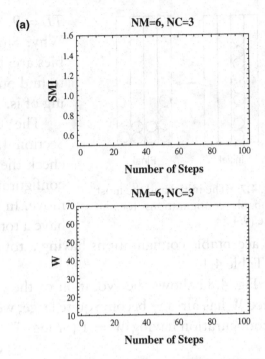

(a)

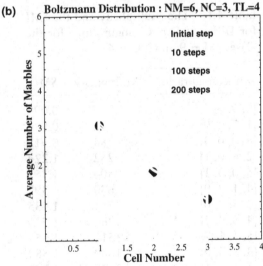

(b)

Fig. 4.6   (a) The evolution of the system shown in Fig. 4.5, with total string length 4 and (b) are the theoretical values of the distribution.[1]

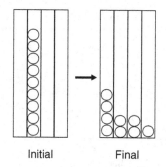

Initial   Final

Fig. 4.7   The initial and the most probable configurations for the case (9;4).

$TL = 9$. Note that this configuration has $SMI = 0$. Why? Simply because we know where all the marbles are; therefore, we do not have to ask questions to find out in which cell the marble that I am thinking of is.

The experiment we do is the same as in Section 4.1. We move two marbles at random and check the total length. If it is 9, we accept the new configuration. If not, we reject it and go to the next move. In the first game, $NM = 9$ and $NC = 4$; we have a total of 220 possible configurations, but only 12 acceptable configurations having a total string length of 9 units. These are listed in Table 4.1.

Fig. 4.8a shows the evolution of the system in terms of SMI and W (note that since W has already become quite large, we plot $\log_2 W$ instead of W). We start with a configuration having $W = 1$ (or $\log_2 W = 0$) and $SMI = 0$. The values then climb within a few steps to higher values of W and SMI. On the scale of these plots, we observe quite large fluctuations in the values of both W and SMI.

The total number of dim-configurations is 220, still quite large, and we cannot draw all of these as we did in Fig. 4.3. However, once we select only those dim-configurations which have a fixed total length, we reduce the number of dim-configurations considerably.

Note also the maximal value of SMI and W. This is important. The configuration that has the largest value of W, as well as of SMI, is $\{4, 2, 2, 1\}$, with $W = 3780$ and $SMI \approx 1.84$ (see Table 4.1).

Table 4.1.   The Multiplicity and the SMI for Different Dim-Configurations for the Case $NM = 9$ and $NC = 4$

Dim-configuration	Multiplicity	SMI
{0, 9, 0, 0}	1	0.000
{1, 7, 1, 0}	72	0.986
{6, 0, 0, 3}	84	0.918
{2, 6, 0, 1}	252	1.224
{5, 0, 3, 1}	504	1.351
{4, 1, 4, 0}	630	1.392
{2, 5, 2, 0}	756	1.435
{4, 3, 0, 2}	1260	1.530
{5, 1, 1, 7}	1512	1.657
{3, 3, 3, 0}	1680	1.584
{3, 4, 1, 1}	2520	1.752
{4, 2, 2, 1}	3780	1.836

Figure 4.8b shows the average number of marbles in each cell. The initial configuration was chosen such that all nine marbles are in cell 2, with none in the other four cells. After 10 steps, the density profile is shown in green. After 100 steps and 200 steps, the profile is almost linear. This profile corresponds to the configuration with the highest W and SMI values.

The important conclusion of these experiments, which differs from the one we have seen in Chapter 3, is that the limiting density profile (i.e. the one after many steps) is not uniform. Here, it looks almost linear, with about 3.5 in cell 1 to about 1 in cell 4.

Figures 4.9 and 4.10 show the evolution of the two systems having the same number of marbles ($NM = 9$) but increasing numbers of cells ($NC = 6$ and $NC = 8$).

The curves of SMI and $\log_2 W$ are almost the same as in Fig. 4.8, and the fluctuations are of similar magnitude. Another important point to be noted is that the maximal values of both SMI and $\log_2 W$ are the same in the three games, as shown in Figs. 4.8–4.10. That means that the maximal values of W and SMI do not depend on the number of cells.

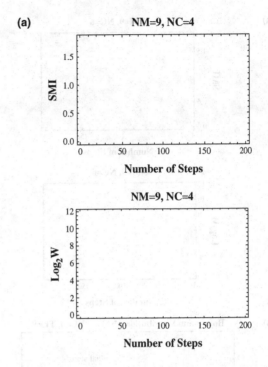

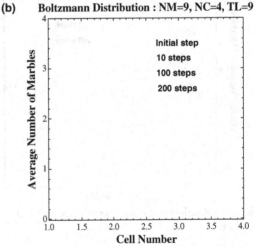

Fig. 4.8 Results for the system with $NM = 9$ and $NC = 4$.

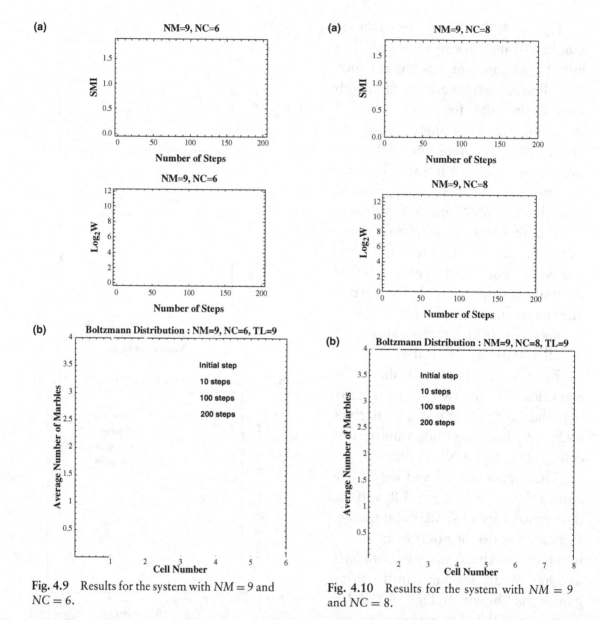

Fig. 4.9   Results for the system with $NM = 9$ and $NC = 6$.

Fig. 4.10   Results for the system with $NM = 9$ and $NC = 8$.

Figure 4.11 shows the density profiles for the three cases. As you can see, the limiting curves for the last two games are almost the same. Furthermore, the limiting density profile is not linear as in the case of $NM = 9$ and $NC = 4$, but there is a clear-cut curvature which is characteristic of the so-called Boltzmann distribution. This is a very important discovery. What we have found is that when we do the experiment with the additional constraint of constant value of the total length (here $TL = 9$), the limiting density profile is *not uniform* but is a steeply declining function of the cell number; the longer the string attached to the marble in the cell, the lower the occupation number in this cell. Clearly, the "ground zero" cell (i.e. the cell with zero string length) will have the maximal occupancy.

Note also that the two curves corresponding to the cases $NC = 6$ and $NC = 8$ are nearly identical in the region between cell 1 and cell 6. Beyond cell 6, the average number of marbles is lower than 0.25, which is

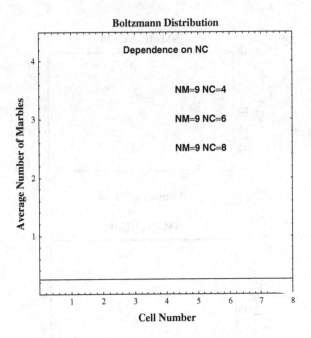

Fig. 4.11   The limiting average number of marbles in each cell, for the three games shown in Figs. 4.8–4.10.

insignificant. We can increase the number of cells as much as we want to — the curves will be unchanged. This is in stark contrast with the results in Chapter 3, where we saw that if we increase the number of cells while keeping the number of marbles *fixed*, we got a uniform distribution over the *entire* range of cells; the larger the number of cells, the lower the average number of marbles per cell. On the other hand, in the experiment done in this section, increasing the number of cells even to infinity would not change the average number of marbles in the first six or seven cells.

4.3. Ten Cells and Variable Number of Marbles

The following experiments are designed to examine the evolution of a system with a fixed number of cells ($NC = 10$) but varying number of marbles ($NM = 10, 30$, and 50), with the total length per marble equal to 3.

Figure 4.12 shows the values of $\log_2 W$ and SMI. Note that in all of the curves, we start with $W = 1$ or, equivalently, $\log_2 W = 0$, and the curve rises sharply to

(a)

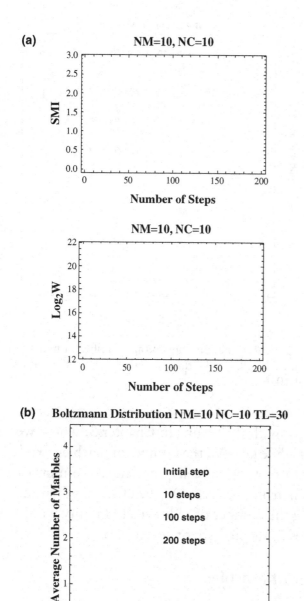

Fig. 4.12   Results for the system with $NM = 10$ and $NC = 10$.

values of the order of 2^{40} to 2^{70} and more. Once the values of W hit the "roof" of the graph, they fluctuate within a narrow range. Note however that the *absolute* range of the fluctuations increases with NM. However, we are interested only in the *relative* fluctuations, and as you can see from Figs. 4.13 and 4.14, these become quite small.

Figures 4.13 and 4.14 show similar results for the cases $NM = 30$, $NM = 50$, and $NC = 10$ (note that the TL changes from 30 to 90 and 150, but the average length per marble is the same as in Fig. 4.12). Note again the steep initial ascent of all the SMI curves, which reach the top with fluctuations at some maximal value around $SMI = 3$. Interestingly, the larger the number of marbles, the larger the maximal value of W. On the other hand, the maximal value of SMI is nearly unchanged and remains about $SMI = 3$. Can you explain why?

Let us look at the density profiles for these experiments. As you can see in Fig. 4.13, the limiting profile has the typical Boltzmann curvature. Of course, the absolute average number of marbles changes when we increase NM, simply because there are more marbles. However, the overall *shape* of the curve does not change much.

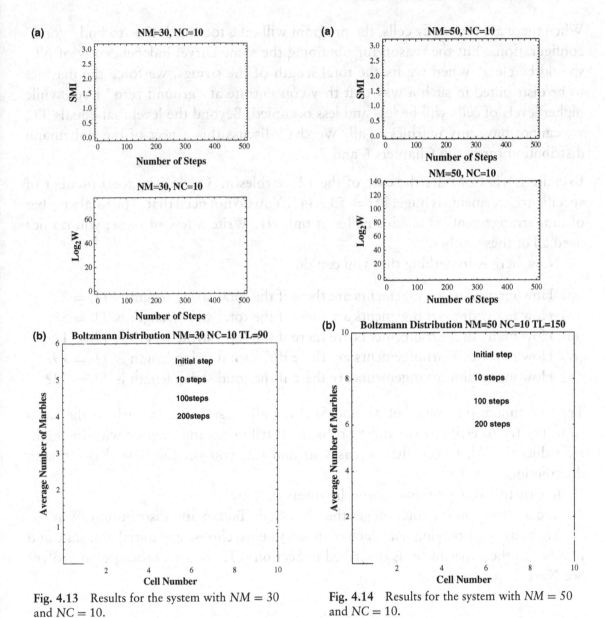

Fig. 4.13 Results for the system with $NM = 30$ and $NC = 10$.

Fig. 4.14 Results for the system with $NM = 50$ and $NC = 10$.

You can also run an experiment with a fixed number of marbles — say, $NM = 50$ — but increase the number of cells — say, to $NC = 20$. You will get the *same* limiting curves no matter how large the NC. We have already seen this independence on NC in the previous section. Here, it takes quite a while to run such an experiment.

When there are too many cells, the program will take too much time to find "good" configurations. But the reason for obtaining the same curve, independently of NC, should be clear. When we fix the total length of the strings, we force the marbles to be distributed in such a way that they concentrate at "ground zero" level, while higher levels of cells will be less and less occupied. Beyond the level that equals TL, we cannot have any marbles at all. We shall discuss this aspect of the Boltzmann distribution further in Chapters 6 and 7.

Exercise E4.1: Consider the case of the 12 marbles in 3 cells. The total number of specific arrangements is huge: $3^{12} = 53,1441$. You do not need that. The total number of dim-arrangements is much smaller at only 91. Write a few of these; you do not need all of these, either.

   Now, here is something that you can do.

 (i) How many dim-arrangements are there if the total string length is $TL = 2$?
 (ii) How many dim-arrangements are there if the total string length is $TL = 5$?
(iii) How many dim-arrangements are there if the total string length is $TL = 12$?
(iv) How many dim-arrangements are there if the total string length is $TL = 19$?
 (v) How many dim-arrangements are there if the total string length is $TL = 22$?

Try to estimate the values of W and SMI for all these cases. In each of the cases (i) to (v), try to estimate the most probable distribution and the one with the highest value of SMI. Check that in cases (i) and (ii), you get the typical Boltzmann distribution.

   In case (iii) you get the uniform distribution. Why?

   In cases (iv) and (v) you will get the "inverted" Boltzmann distribution. Why?

   Try to imagine playing with each of these systems; choose any initial dim state and proceed to the experiment as described in Section 4.1. How will the system evolve? See Note 2 and Table 4.2.

4.4. Some Simple Processes Involving Transferring of Marbles

In this section, we shall perform some simple experiments, similar to those we have done in Section 3.3, but now we shall have the strings attached to the marbles.

4.4.1. *Expansion*

Figure 4.15 shows a process of "expansion" similar to the process described in Section 3.3 (Fig. 3.21). We start with an equilibrated system of $NM = 9$ and $NC = 4$, then add to this system another system of four empty cells. Thus, in the new system we have $NM = 9$ and $NC = 8$. One would expect that marbles will "spread" in the entirely new *expanded* system: This does not occur. The distribution remains unchanged in spite of the increase in the "volume" of the system. This is in sharp contrast to the result of the expansion process described in Section 3.3 and Fig. 3.21. I urge you to ponder on the reason for these different results.

Table 4.2. All Possible Dim-Configurations for the 12;3 Game, for Different Values of *TL*

Dim distribution		W	SMI
$TL = 2$	(10, 2, 0)	66	0.650
	(11, 0, 1)	12	0.414
$TL = 5$	(7, 5, 0)	792	0.979
	(8, 3, 1)	1980	1.188
	(9, 1, 2)	660	1.041
$TL = 12$	(0, 12, 0)	1	0.
	(1, 10, 1)	132	0.816
	(2, 8, 2)	2970	1.251
	(3, 6, 3)	18480	1.5
	(4, 4, 4)	34650	1.585
	(5, 2, 5)	16632	1.483
	(6, 0, 6)	924	1.
$TL = 19$	(0, 5, 7)	792	0.980
	(1, 3, 8)	1980	1.189
	(2, 1, 9)	660	1.041
$TL = 22$	(0, 2, 10)	66	0.650
	(1, 0, 11)	12	0.414

4.4.2. *Pure Mixing*

Figure 4.16 shows the process of pure mixing. Here, we start with 9 blue marbles in 4 cells, and another 9 red marbles in 4 cells, and combine them into one system of 18 marbles in 4 cells. We have already seen the evolution of the system of (9;4) in Section 4.1. Here, all we have to do is to start with the equilibrated systems of pure blues and pure reds, and find a new equilibrium state for the entire mixture of 18 mixed marbles in 4 cells. The change in *SMI* is zero in this case, as it was in the case of *pure* mixing discussed in Chapter 3 (note that the total string length is conserved).

4.4.3. *Pure Assimilation*

This is essentially the same as the pure assimilation discussed in Chapter 3. Again, we start with two initially equilibrated systems of (9;4) and we combine them into

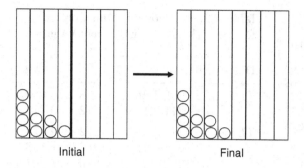

Fig. 4.15 The process of "expansion" from four cells into eight cells. When the barrier between the two systems is removed, the distribution of marbles does not change.

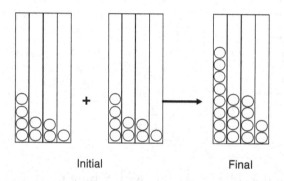

Fig. 4.16 A pure mixing process.

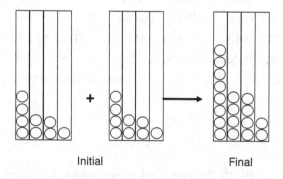

Fig. 4.17 A pure assimilation process.

one system of (18;4) (Fig. 4.17). As in the case of pure assimilation in Chapter 3, the value of the *SMI decreases* in this process. The final distribution of the marbles has roughly the same form but in each level there are twice as many marbles as in the initial separated systems (note that the total string length is conserved).

It is now clear that in any process carried out in a constant-string-length system, the final equilibrium configuration can be predicted easily without actually doing the experiments. There is nothing unpredictable or interesting in these experiments. We shall discuss this topic further in Chapters 6 and 7. However, there is one very important process in the string-attached-systems, which we shall discuss in the following sections.

4.5. Dependence of *SMI* and *W* on the String Length *TL*

In this chapter, we have seen that simple processes such as expansion, mixing, and so on do not provide anything new that we did not see in Chapter 3. However, there is a class of processes involving Boltzmann distributed marbles that is both new and interesting, and its real-life counterpart is central to the understanding of the Second Law. This class includes processes in which strings are *exchanged* between two systems.

Before doing some of these experiments, we will first study the dependence of the density distribution on the total string length.

Let us examine a simple example. The system consists of 12 marbles in 3 cells (12;3). Clearly, if the total string length is zero, then all the marbles must be in ground zero and no amount of shaking of the system can change that. Next, when the total string length is 1, there exists only one dim-configuration that fulfills the constraint $TL = 1$; that is, 11 marbles in cell 1 and 1 marble in cell 2. Again, no matter how much we shake this system, the density distribution will remain unchanged.

With $TL = 2$ we have two possible configurations — {10, 2, 0} and {11, 0, 1} — with multiplicities of 66 and 12, respectively. It is easy to check that with $TL = 3$ we again have again two configurations — {9, 3, 0} and {10, 1, 1} — with multiplicities of 220 and 132, respectively. Thus, for each TL we can write down all the possible configurations and estimate their multiplicities.

Table 4.2 shows all the possible configurations for some values of TL. The density profiles for this system with varying TL are shown in Fig. 4.18.

When we increase TL to, say, $TL = 5$, the curve becomes shallower. Clearly, with string length of $TL = 5$, the system can spread its marbles further to the right (see Table 4.2). Note that the configuration {8, 3, 1} has the largest multiplicity value $W = 1980$, as well as the largest value of $SMI = 1.1887$. Why? If you were to play the 20Q

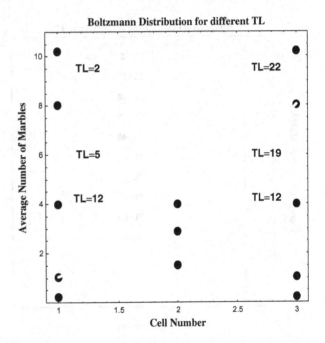

Fig. 4.18 Dependence of the density distribution on TL for the game 12;3.

game on this system, and you know the probability distribution in this distribution, it will be relatively more difficult to find the hidden marble.

Let us move to $TL = 12$. As we see in Fig. 4.18, the density profile is flat; that is, the equilibrium distribution is uniform. Look at Table 4.2 and you will find that there are seven possible configurations that fulfill the condition $TL = 12$. The one with the largest value of W as well as SMI is the uniform distribution {4, 4, 4}. The significance of this case will be discussed further in Chapters 6 and 7. At this stage, all you have to understand is that the larger the value of TL, the more difficult it will be to play the 20Q game on this system, up to a point (here $TL = 12$).

Once the TL is larger than 12, the curves become mirror images of the curves with TL smaller than 12. We are not interested in these cases here. I shall give you a quick hint why, just to satisfy your curiosity. The case $TL = 0$ corresponds to the absolute zero temperature; all marbles are in the ground state. $TL = 12$ corresponds to infinite temperature, at which the distribution becomes uniform. The curves for TL larger than 12 correspond to "negative temperatures," and therefore will be of no interest to us.

Figure 4.19 shows the density distribution for the system (20;10) with varying string lengths. Each of the curves was obtained by running the same experiment as in Sections 4.2 and 4.3; only the values of TL were changed from one experiment to the other. In each experiment, we also followed the evolution of the SMI and the multiplicity, W. The evolution of these systems is quite similar to the ones we had observed in Sections 4.2 and 4.3, and will not be discussed here.

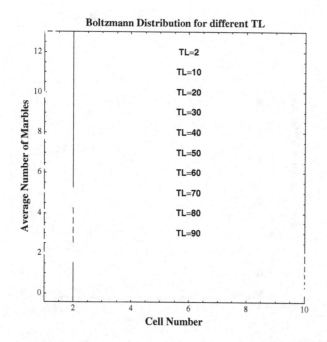

Fig. 4.19 Density distributions for different TL for the game of 20;10.

For each value of *TL* we record the maximum (or the equilibrium) value of the *SMI* and *W*. A list of these values is given in Table 4.3. Note that each row in the table is the result of an entire experiment on the (20;10) system with a specific value of *TL*.

A glance at this table shows that the larger the value of the *TL* (between 1 and 90), the larger the *equilibrium* value of *SMI*, as well as the corresponding value of $\log_2 W/NM$. This is a very important result! Keep that in mind; we shall find in Chapters 6 and 7 that this result is at the heart of the Second Law.

Aside from the dependence of *SMI* and W on *TL* that can be read from Table 4.3, we can discover something hidden in the table if we plot the values of *SMI* and $\log_2 W/NM$ for the different values of *TL*.

Figure 4.20 shows the maximal (the equilibrium) value of *SMI* and $(\log_2 W)/NM$ obtained from each experiment with different values of *TL*. Clearly, we see that both *SMI* and $\log_2 W/NM$ increase with the increase in *TL*. This is what we already have read from Table 4.3. However, from this figure we can see something more, something that we could not have easily seen from the table. Can you tell what it is?

Table 4.3. Equilibrium values of *SMI* and $\log_2 W/NM$ for the system 20;10, for different *TL*.

TL	SMI	$\log_2 W/NM$
1	0.290	0.210
5	0.902	0.707
9	1.296	1.019
13	1.596	1.251
17	1.841	1.436
21	2.047	1.590
25	2.225	1.723
29	2.382	1.838
33	2.519	1.940
37	2.642	2.032
41	2.750	2.113
45	2.847	2.186
49	2.933	2.252
53	3.009	2.310
57	3.075	2.361
61	3.133	2.406
65	3.182	2.444
69	3.224	2.477
73	3.258	2.503
77	3.285	2.524
81	3.304	2.540
85	3.316	2.549
89	3.322	2.554

I will let you examine Fig. 4.20 for yourself, and let you "discover" what is so conspicuous in the figure that is not easily revealed by the table. If you find out the answer to this question, you will also learn an important lesson in doing research. Sometimes plotting a list of numbers reveals much more than just reading it from a table. In the next section, we shall see the implications of this particular *shape* of the curves in Fig. 4.20, for some process involving "exchange of strings" between two systems. The relevance to the Second Law in real processes is deferred to Chapter 7.

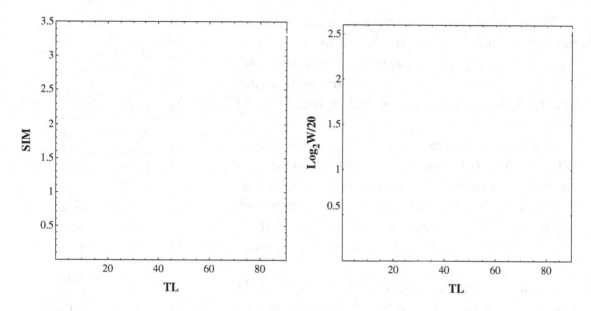

Fig. 4.20   Plot of *SMI* and $\log_2 W/20$ as a function of *TL* for the game of 20;10.

4.6. A Process Involving Exchange of String Between Two Systems

The following experiment was carried out on the same game as the one discussed at the end of Section 4.5; that is, the system of 20 marbles in 10 cells.

We start with two systems; one system has the *TL* = 60, and the second has the value of *TL* = 20. The experiment we shall do is simple. We *transfer* 10 units of string from the higher *TL* value system to the lower *TL* value system. Thus, after the transfer we have two new systems, one with *TL* = 50 and one with *TL* = 30. This process is described in Fig. 4.21.

Note that we began with two *equilibrated* systems. This means that each of the two systems has already reached its equilibrium state. After the transfer of a piece of string of 10 units of length, we get a new pair of systems, one with a higher value of *TL* and one with a lower value of *TL*. This pair of new systems is not at equilibrium. Therefore, after the transfer has been done, we do the equilibration of the two systems. This simply means that we "shake" each of the system until we reach a new equilibrium state for each of the two systems. What will happen?

If you have followed the discussions in the previous section, you will realize that we do not need to do the experiment of equilibration. We have already done that for each value of *TL*, and the results are shown in Figs. 4.19 and 4.20.

Thus, we know that as the value of *TL* increases, the distribution becomes flattened. This means that when 10 units of string are transferred, one system will have a steeper distribution while the other will have a flatter distribution. The arrows in Fig. 4.22 show the direction of change of the two distributions.

As you can see from Fig. 4.22, there is nothing unexpected in the change of the density profiles of the two systems. This result can be predicted without actually performing any experiment.

We can also predict the change in the values of the *SMI* and of *W* as a result of the transfer of string. We have already recorded the dependence of *SMI* and *W* on *TL* in Fig. 4.20. All we have to do is

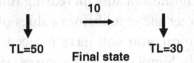

Fig. 4.21   The process of exchange of strings between two systems from the higher to the lower value *TL*.

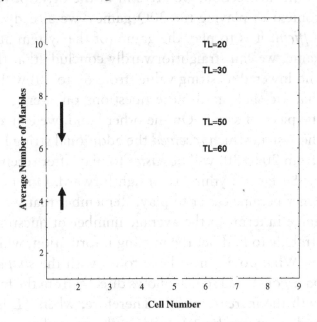

Fig. 4.22   The change in the density distributions of the two systems after transferring 20 units of length and equilibrating the system.

read the values of *SMI* corresponding to the values of $TL = 20$ and 30 for one system, and of $TL = 60$ and 50 for the second system. The same can be done for the W values, simply reading these from the plots of $\log_2 W/NM$ against TL.

Here is the unexpected result. I hope that by now you have examined the *shape* of the curves in Fig. 4.20, and you have noticed that both the *SMI* and the $\log_2 W/NM$ curves are monotonically increasing functions of TL. In plain language, the larger the TL, the larger the equilibrium values of *SMI* and of $\log_2 W/NM$.

In addition, you will have noticed that the two curves in Fig. 4.20 are concave downwards. Simply stated, the curves rise *steeply* at lower values of TL and the slopes become less and less pronounced as TL increases and, eventually, the curve levels off.

This particular *shape* of the two curves is conspicuous in Fig. 4.20, and simple to state. We shall see that this shape also determines the *sign* of the change of *SMI* in this process. In Chapter 7, we shall also see that this is relevant to the Second Law of Thermodynamics.

In this section, we are still in the descriptive phase of our investigation. In the context of playing the 20Q games, we already know that the larger TL, the more *difficult* it is to play the game (of the system at equilibrium). In terms of the 20Q game, we can straightforwardly conclude that the system which *donated* the string and lowered its string value (from 60 to 50) will now be *harder* to play (in the sense that we shall need more questions on average to find out), than before shedding the piece of string. On the other hand, we can also straightforwardly conclude that the system which *received* the additional string length and has increased its TL value (from 20 to 30) will be *easier* to play after receiving additional string.

So far everything is straightforward. One game became harder to play, and the other became easier to play. Remember that we agreed to measure the "size" of the game in terms of the average number of questions one needs to ask in the smartest strategy to find out the missing information, which we have denoted as *SMI*.

What do all these have to do with the *shape* of the curves in Fig. 4.20? All we have concluded so far follows directly from the fact that *SMI* monotonically *increases* with the increase of TL. Therefore, when TL increases by 10 units in one system and decreases by 10 units in the second system, the first becomes more *difficult* to play (higher value of *SMI*) and the second becomes *easier* to play (lower value of *SMI*).

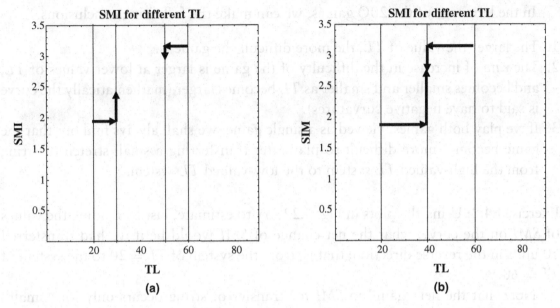

Fig. 4.23 (a) The change in the *SMI* for transferring of 10 units of string length. (b) The change in the *SMI* for the *spontaneous* exchange of string.

However, here comes the most astonishing discovery you will ever make. Look carefully at Fig. 4.23a and watch what happens when we increase the value of *TL* from 20 to 30. The value of *SMI* increases by about 0.42. On the other hand, when the value of *TL* decreases from 60 to 50, the value of the *SMI* decreases by about 0.17 (these numbers were read from the curves and are indicated by the arrows in Fig. 4.23a).

This means that in the process of transferring 10 units of length, the total length of the strings in the entire system is *conserved*: We had a total of $60 + 20 = 80$ units before the transfer, and 80 units after the transfer. The value of *SMI* is however *not conserved* in the process. It has increased by about 0.25 units on the *SMI* scale.

One can prove mathematically that this "gain" in *SMI* will occur for any *small* transfer of string length from one system to another, and that this result is related to the *curvature* of the graph of *SMI* against *TL*. This is the reason why I urged you to examine the *shape* of the curves of *SMI* and see what cannot be seen from looking at the numbers in Table 4.3.

In the language of the 20Q games, we can make the following conclusions:

1. The larger the value of *TL*, the more difficult the game is.
2. The *rate* of increase in the difficulty of the game is larger at lower values of *TL*, and becomes smaller and smaller as *TL* becomes larger (mathematically the curve is said to have negative curvatures).
3. If we play both games, viewed as a single game, we shall always find out that the game becomes more difficult to play after transferring a small stretch of string from the high-valued *TL* system to the low-valued *TL* system.

Exercise E4.2: Using the plots in Fig. 4.22, try to estimate, just by reading the values of *SMI* on the curve, what the net change of *SMI* would be if we had transferred 10 units in the reverse direction; that is, from the system of *TL* = 20 to the system of *TL* = 60.

Note that the net "gain" in *SMI* for transfer of string occurs only for "small" values of the length of the string. How small is "small" will become clear in the next section.

Exercise E4.3: Try to estimate the form of the distribution of the outcomes of a die if you are given the averages of outcomes:

Average: 2, 3, 3.5, 4, 5.

What is so special about the average 3.5?

4.7. A Spontaneous Transfer of String from the High-Value to a Lower-Value *TL*

Before moving on to the next experiment, it is of utmost importance that you understand all the details of the experiment that we have performed in Section 4.6.

Each of the two systems consists of 20 marbles distributed among 10 cells. "Shaking" of the system means that we allow randomly chosen marbles to jump from their cell into another randomly chosen cell. In each step, the total string length of *each single system* must be conserved. This means that if one marble goes to the right, one or more marbles must go to the left.

Clearly, not every move of a marble to a new cell is accepted. For instance, a marble in cell 2 moving to cell 4 will need two units of string length in order to do so. This step in itself will not be accepted. However, if another marble goes from cell 6 to cell 4, it can "donate" the two units of length to the first marble. Therefore, the combined moves of the two marbles will be accepted.

Another accepted move is that one marble goes from cell 2 to cell 4 as before, and two other marbles go one cell to the left — say, from cell 5 to cell 4, and from cell 7 to cell 6. The combined moves of the three marbles will also be accepted.

We can imagine that the string attached to the marble can be *exchanged* among the marbles. Any time one marble moves towards the right it needs additional string, which it gets from one or more marbles moving to the left, so that the total string length is conserved.

In all the experiments in Sections 4.2 and 4.3, we started with some initial configuration and made many moves of marbles among the cells, but we have *accepted* only those moves that are consistent with the requirement that the total string length is conserved.

For each configuration we can play the 20Q game (I think of a specific marble and you have to find out in which cell that specific marble is). Therefore, each newly accepted move creates a new game, characterized by a new *SMI*, and also by a new *W* (the number of specific configurations pertaining to the dim-configuration).

In all the experiments we carried out in Sections 4.2 and 4.3, we saw that both *SMI* and *W* start from a relatively small value, then increase as the number of (accepted) steps increases, and eventually reach a leveled-off state. When the number of marbles is quite large, the fluctuations about this limiting level are quite small, and we call this limiting game the equilibrated game.

In the language of the 20Q game, we can say that whenever we start with a relatively "easy" game (requiring a small number of questions), as the system evolves toward the equilibrium level it always evolves from a relatively "easy" game to a more "difficult" game. For each game with a given value of *TL*, the system evolves differently. The *limiting* value of *SMI* is different for different values of *TL*. These values are shown in Fig. 4.20.

Remember that in Fig. 4.20 we recorded the *limiting* values of *SMI*; that is, the maximal value of the system having a specific value of *TL*, reached after many steps.

From Fig. 4.20, we can conclude that the limiting game with larger *TL* is more difficult to play than the game with a lower *TL*.

The process we carried out in Section 4.6 was as follows. Each of the two systems with $TL = 20$ and $TL = 60$ was first equilibrated *separately*. That means that the strings attached to the marbles in each system could be exchanged among the 20 marbles of the *same system*. We next transferred 10 units of string length from one system to the other. Once we had done that, we allowed the two systems (that can now be referred to as the $TL = 30$ and $TL = 50$ systems) to evolve *separately* to their new equilibrium level. These new equilibrated systems have new equilibrium values of *SMI*, as recorded in Fig. 4.23.

The important thing to remember is that the strings of each system *can be exchanged among the marbles within the same system, and not between marbles belonging to different systems.*

In the experiment in Section 4.6, we have, by our own free will, free from coercion or intimidation, transferred 10 units from one system to another. The exchange of string length was not done *by* the systems themselves, but was imposed by us *on* the systems from the "outside."

Now, for the next, most stunning experiment, watch carefully what we do and what we do not do. We start with the two systems initially at the same states as in Section 4.6. Again, as before, each of the two systems (the $TL = 20$ and $TL = 60$) has evolved into its own equilibrium level *separately*.

Unlike the experiment in Section 4.6, we *do not* transfer any strings from one system to the other. Instead, we relax the requirements that the strings can be exchanged only among marbles belonging to the *same* system, and allow all the 40 marbles of the *combined* system to exchange strings as they "wish."

Please pause and read the last paragraph again. Make sure you understand the difference between this and the previous experiment. Note that we still do not permit the transfer of *marbles* from one system to another as we did in the experiments in Section 4.6. Each of the systems preserves its own 20 marbles. What we have changed in this experiment is to lift the rule that string length must be conserved in each system. Instead, we *allow string to flow freely between the two systems, keeping the conservation rule for the combined system.*

We can imagine a barrier separating the two systems which precludes any transfer of strings between the two systems. The experiment we do is simply to remove this barrier and allow free flow of strings between the two systems, keeping the total length of the string in the *combined* system fixed (here, *TL* (combined) = 20 + 60 = 80).

A quick note for the reader who is impatient and is bored by these strange experiments, which no one has ever done: The strings stand for the energies of the particles. The barrier is an adiabatic wall; that is, a wall between two real systems which prevents heat transfer from one system to another. "Removing the barrier" means replacing the adiabatic wall with a diathermal, or heat-conducting wall. What we are going to observe is a spontaneous process for which the total string is *conserved*, but the *SMI* is *not conserved*. This sounds like a marbled-echo of the Second Law. For more on that, see Chapter 7.

Now, after lifting the barrier, we shake the *combined* system and let it evolve to a new equilibrium level. What do you expect to happen?

If you had never heard about the Second Law of Thermodynamics you would not be able to "predict" what will happen. However, if you know something about the Second Law, you might be able to "predict" what will happen. You will probably say that if we allow the two systems to exchange strings, then the string will flow from the high *TL* system to the low *TL* system. That is a correct prediction. However, I suspect that you have arrived at the correct prediction but for the wrong reason. Wait a minute. I will explain that soon, after we do the experiment.

Let us carry out the experiment, allowing exchange of strings between the two systems, and follow the co-evolution of the pair of systems as depicted in Fig. 4.24. Figure 4.25 shows that the density profiles of the two systems will *approach* each other. The final distribution will be an exact intermedial one; the high *TL* system will evolve from *TL* = 60 to *TL* = 40; the low *TL* system will evolve from *TL* = 20 to *TL* = 40. Furthermore, we can see in Fig. 4.23b how the *SMI* of each system will change. At equilibrium, each of the systems will end up having the same value of *TL* (here *TL* = 40), and the same value of *SMI* = 2.72 (read from the plot in Fig. 4.23). This is the equilibrium value of *SMI* for a system with *TL* = 40.

Now, the more difficult question: Why has the system evolved to this particular state? Clearly, we *did not* transfer any strings from one system to the

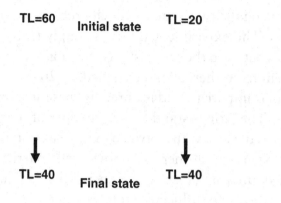

Fig. 4.24   Schematic process of the spontaneous exchange of string.

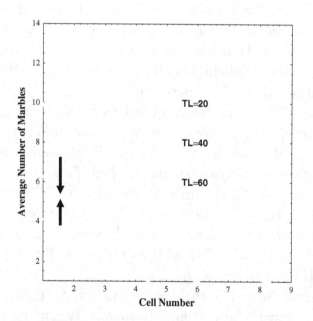

Fig. 4.25   The change in the density distribution for the spontaneous process in Fig. 4.24.

other, as we did in the experiment of Section 4.6. Here, the systems *themselves* "decided" to do that voluntarily! The high *TL* system to donate 20 units of string, and the low *TL* system happily accepted the donation. So my "why" question has two parts: Why have the systems exchanged strings in the first place? Why did they exchange exactly that amount; no more than 20 units, no less than 20 units?

If you have never heard of the Second Law you will surely be stunned by watching the co-evolution of the system. You will have no idea why the co-evolution of the two systems took that particular path. If, on the other hand, you have heard about the Second Law, you might hastily construct an answer to my questions as follows:

> "That is clear. I know that the *SMI* is a disguised entropy and that *TL* represents the temperature of the system.

As we have seen in the experiment in Section 4.6, when strings flow from the high *TL* to the low *TL*, the entropy (disguised as *SMI*) has increased. Therefore, in the present experiment also the entropy (disguised as *SMI*) has increased, as can be read from Fig. 4.23b. The Second Law states that in any spontaneous process in an isolated system (here represented by the combined systems), the entropy always increases".

What you have said is *almost* a perfectly correct answer. Unfortunately, it is not the answer to my question. What you said is indeed correct. You have correctly described *what* has happened. However, you did not answer the question of *why* it happened in this particular way.

You correctly referred to Fig. 4.23 to see that *what* happened in the process is that the net *SMI* has *increased* in the process, but the answer to the question of *why* lies not in Fig. 4.23 but in Fig. 4.26. As in Fig. 4.23, we show in Fig. 4.26 the changes in $\log_2 W/20$ for the two processes of transferring of string. You can read from the figure that in both experiments, the net change in $\log_2 W/NM$ is *positive*. Higher $\log_2 W$ values mean higher probability.

At this point, we can also answer the question we raised in Section 4.6. How "small" must the transfer of string be in order to have a net increase in *SMI*? The

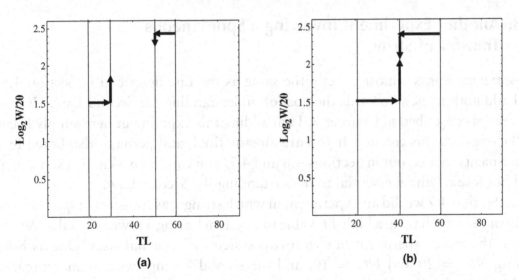

Fig. 4.26 (a) The changes in the $\log_2 W/20$ for transferring of 10 units of string length. (b) The change in $\log_2 W/20$ for the spontaneous exchange of string.

answer is not larger than 20 units for this particular case. Beyond 20 units, the *SMI* will start to decrease.

We shall discuss these questions (and answers) in Chapters 6 and 7. We are now only in the descriptive stage of our work. But to give you a quick hint, let me just say this. Do not mix up cause and effect. What *drives* the system, any system, is not the *entropy* but the *probability*. The probability (disguised as *W*) is the *cause* of the specific direction in which the system evolved spontaneously. The entropy (disguised as *SMI*) is only one of the *results* of the evolution of the system. And besides, I did not want to spoil your excitement about the explanation of the game you offered above. You were correct about the entropy disguised here as *SMI*, but you were not correct in identifying *TL* with temperature.

Therefore, be patient until we get to Chapters 6 and 7. For now, we have to do some more experiments in Chapter 5, the outcomes of which are not less exciting than the ones we have done in this chapter. Meanwhile, we add here one more experiment of the same kind as described above but with an important "moral". You will find out why I added this experiment in Chapter 7.

4.8. Another Experiment Involving a Spontaneous Transfer of String

This experiment is almost exactly the same as the one described in Section 4.6. It will add nothing new towards the goal of understanding the Second Law. However, for reasons described in Chapter 7, I am adding this experiment here simply because it "belongs" to this chapter. If you are already tired, and perhaps also bored by the experiments carried out in Sections 4.6 and 4.7, you can safely skip this section. You will not lose anything essential to understanding the Second Law.

In Section 4.7 we did an experiment in which string was transferred *spontaneously* from one system having a high *TL* value to a system having a lower *TL* value. We now repeat the same experiment but with two systems of *unequal* sizes. One as before, having $NM = 20$ and $NC = 10$, and the second having twice as many marbles, $NM = 40$ and $NC = 10$.

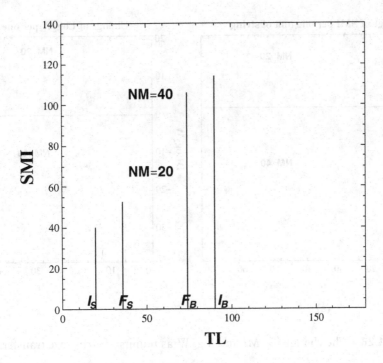

Fig. 4.27 Transfer of string between a "big" and a "small" system.

The experiment is exactly the same as before. We start with the two systems at equilibrium and then we combine them in such a way that the total string length is conserved.

Figure 4.27 shows the *SMI* of the two systems as a function of *TL*. We start the experiment with the small system with $TL = 20$, and the big system with $TL = 90$. These two points are marked by I_S and I_B (for the initial state of the *small* and the *big* systems, respectively) in Fig. 4.27.

Once we start the shaking of the combined systems, we shall see that string will flow from the big to the small system. We can read from Fig. 4.27 how the *SMI* of each system will change. The *SMI* of the small system will *increase* as a result of *gaining* string length, and the *SMI* of the big system will *decrease* as a result of *losing* string length. Fig. 4.28a shows the *change* in the *SMI* of the two systems for transferring n

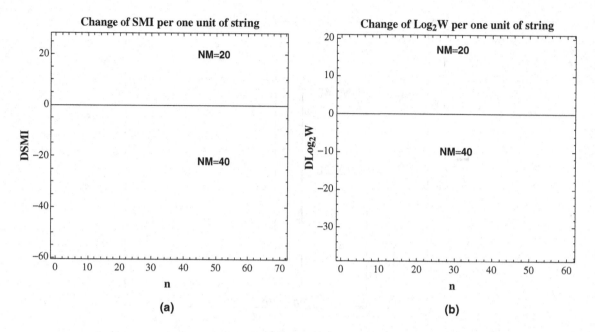

Fig. 4.28   The change in *SMI* and $\log_2 W$ as *n* units of string are transferred.

units of string length between the two systems. The $n = 0$ point corresponds to the initial states of the two systems. As *n* units of length are transferred, the values of *SMI* of the small system increase, and of the big one decrease. This can also be read from Fig. 4.27.

Looking at the overall change in the *SMI* for the two systems in Fig. 4.28a, you might conclude that for any *n*, the decrement of *SMI* of the big system is larger than the increment of *SMI* of the small system. For example, for a transfer of say, 70 units, the *SMI* of the big system decreased by about 60 units, whereas the *SMI* of the small system has increased by about 22 units.

Similar behavior is exhibited by the corresponding $\log_2 W$ curves of the two systems (Fig. 4.28b). From this overall behavior of both *SMI* and $\log_2 W$ you might hastily conclude that no string will flow spontaneously from one system to the other.

However, a more careful examination of the two curves in Fig. 4.28a (as well as in Fig. 4.28b) shows that the "overall" increase of *SMI* (and $\log_2 W$) of the small system is indeed smaller than the decrease of *SMI* of the big system. But if you look

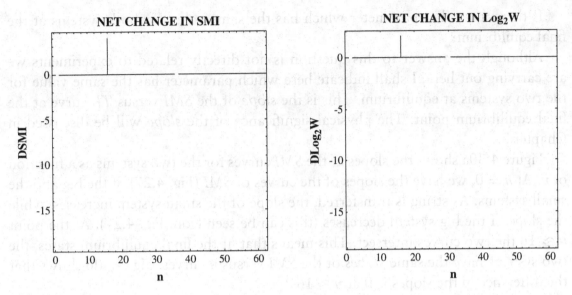

Fig. 4.29 The change in the net *SMI* (denoted *DSMI*) and of $\log_2 W$ (denoted $D \log_2 W$) of the combined systems as we add *n* units of string.

at very *small* values of *n*, you will see that the *increment* in *SMI* (as well as $\log_2 W$) for the *small system* is *larger* than the *decrement* of the *SMI* for the *big system*. This finding suggests that we plot the *net change* of *SMI* of the combined system as *n* units are exchanged between the two systems. Figure 4.29a shows the net change in *SMI* and Fig. 4.29b shows the net change in $\log_2 W$ for the combined system. As is clear from the figure, at about $n \approx 16$ units, the net change in *SMI* (and in $\log_2 W$) is *positive* and maximum. This means that the most probable amount of string that will be transferred between the two systems is about $n \approx 16$.

Going back to Fig. 4.27, we have denoted by F_B and F_S the final states of the two systems after 16 units have been transferred from the big to the small system.

This plot is quite different from the one in Fig. 2.23b, where the final equilibrium point was exactly halfway between $TL = 20$ and $TL = 60$ (i.e. the final point was $TL = (60 + 20)/2 = 40$).

Thus, we see that in the process of spontaneous transfer of string, the final equilibrium state is not the one for which the *TL* values of the two systems are equal.

Is there any other parameter which has the same value for both systems at the final equilibrium?

Although the answer to this question is not directly related to experiments we are carrying out here, I shall indicate here which parameter has the same value for the two systems at equilibrium. This is the *slope* of the *SMI* versus *TL* curve at the final equilibrium point. The physical significance of the *slope* will be discussed in Chapter 7.

Figure 4.30a shows the slopes of the *SMI* curves for the two systems as a function of *n*. At $n = 0$, we have the slopes of the curves of *SMI* (Fig. 4.27) of the big and the small systems. As string is transferred, the slope of the small system increases, while the slope of the big system decreases (this can be seen from Fig. 4.27). At the point $n \approx 16$ the two curves intersect. This means that at the final equilibrium states, the two systems have the same *slopes* of the *SMI* versus *n* curves. Fig. 4.30b shows that the difference in the slopes is 0 at $n \approx 16$.[3]

Thus, in general, when we combine two systems of different sizes there will be spontaneous transfer of string from the system with *smaller* slope of the *SMI* curve

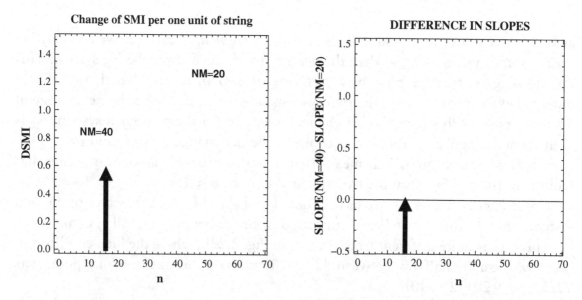

Fig. 4.30 (a) Plot of the slopes of the two curves in Fig. 4.27 as *n* units of string are transferred from the "big" to the "small" system. (b) The difference in the slopes in (a).

(as well as the $\log_2 W$ curve) to the system with *larger* slope. The transfer of string will stop at that point when the slopes of the two systems will be the same.[3]

Exercise E4.4: A most challenging exercise.

Is it possible that a "small" system (i.e. a game having a smaller number of marbles, smaller *TL*, and smaller *SMI*) will transfer string spontaneously to the bigger system?

Note that in the two processes discussed in Sections 4.7 and 4.8, string flew spontaneously from the high-*TL* system to the low-*TL* system. It was relatively easy to find a specific example of this kind of process. It is less easy to find an example where string will flow from the *lower* value of *TL* to the *higher* value of *TL*.

This exercise is addressed to a reader who has some knowledge of thermodynamics. To those who do not know any thermodynamics, I suggest skipping this exercise. It is too challenging, and yet does not add anything to understanding the Second Law.

For those who will take the challenge, I suggest that you do a little research work. Check first the meaning of the inverse slopes of the *SMI* versus *TL* curves. Then find out the criterion for a spontaneous flow of string from one system to another. Once you have found this criterion, it will be easy to find an example of two systems that meet this criterion. If you can't find it, look at Note 4.

4.9. What Have We Learned in this Chapter?

In all of the experiments we have done in Chapter 3, the resulting limiting density distribution was always a uniform one. We rationalized that observation by an argument of equivalence. Since all the cells are equivalent, there is no reason for the system to prefer any one particular cell. Of course, we have started with an arbitrary configuration. I chose purposely the one with the least probability. But once we performed a few hundreds of steps, and the system had enough time to randomize, there was no reason that one cell would on average receive more marbles than the other. Thus, we have arrived at the uniform distribution. In fact we could have guessed that the uniform distribution *must* be the eventual outcome of the experiment. The argument is essentially the same as the one that we used to "guess" that the probability of obtaining any specific result — say, "4" in throwing a die — is 1/6.

If our reasoning is so compelling for the experiments of Chapter 3, why in this chapter did we get a completely different result? Why did the attachment of the strings and the requirement of conservation of the total length have such a drastic and unexpected result?

Qualitatively, the answer is simple. Because of the attached strings, we cannot hold on to the argument of equivalence of the cells. Each cell is characterized by a different length of a string, and this in itself breaks the equivalence. Furthermore, we can rationalize the appearance of a monotonic decreasing distribution as follows.

Suppose we start with all marbles in the second cell. (If we choose the initial configuration to be such that all marbles are in either the first or the last cell, then the system will not evolve at all. Why?) Let us assume that we started with all marbles in cell 1 (as in Fig. 4.31a), and start shaking the system. Since we required the conservation of the total string length, at least two marbles must move at any step. Clearly, if *one* marble moves *one* cell to the right, another *one* must move *one* cell to the left. This will conserve the total length, and we will have a new configuration like the ones in Fig. 4.31b.

On the other hand, if *one* marble moves *two* cells to the right, then *two* marbles must move *one* cell to the left (Fig. 4.31c). If *one* marble jumped *three* cells to the right, *three* marbles must be moved *one* cell to the left, and so forth. Thus, for any move of *one* marble x cells to the right, x marbles must be moved *one* cell to the left.

The net effect is that as you try to spread the marbles longitudinally to the right, accumulation of marbles on the left will be faster. This explains why the asymmetry of

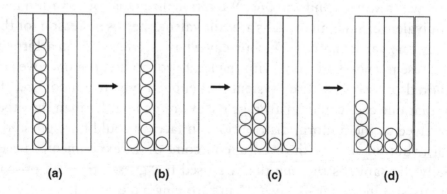

(a)   (b)   (c)   (d)

Fig. 4.31   The building up of the Boltzmann distribution.

the distribution in favor of larger population at ground zero, and gradually declining population as we go farther to the right. The final distribution looks like the one in Fig. 4.31d.

All of the abovementioned arguments are qualitatively correct, but they do not tell us why this particular shape of the decreasing function is obtained. For this, there is no way to find out without any mathematics. We shall discuss this question further in Chapter 6.

The Boltzmann distribution is important, but not essential for understanding the Second Law. It is extremely important, however, for physics in general, and thermodynamics in particular. All the distributions of molecules in their energy levels (electronic, vibrational, rotational, and translational) are governed by the Boltzmann distribution. If this is way beyond you, think of the distribution of the density of the molecules above you. It is also governed by the Boltzmann distribution. Thanks to Boltzmann, pilots can measure their altitude by measuring the density (or the pressure) of the air in whichever location they are without sending any signal down to earth. This is indeed a remarkable achievement.

We shall discuss the motional "energy levels" in the next chapter, where we shall discover the Maxwell–Boltzmann distribution.

Snack: How to Choose Between Two Equal-Probability Alternatives

Many years ago, I attended a lecture on game theory which was given by the late Michael Maschler, a mathematics professor at the Hebrew University in Jerusalem. At that time, Maschler quoted Kenneth Arrow, the Economics Nobel Laureate (1972) as the source of the story. In preparing this book, I wrote to Arrow to verify the details but all he could say was that he heard the story from Merrill Flood, who was an Operations Research Officer with the United States Air Force in the Pacific during WWII. The exact details of the story have not been validated, but the pilots' dilemma and the way the probabilities are subjectively perceived are correct.

This is a true story about a squadron of American bombers assigned to the Pacific Islands during WWII, whose mission was to carry out bombing sorties to an island which belonged to Japan.

A fighter plane's maximum load capacity was, say, three tons. Normally, the plane carried two tons of fuel for a round-trip and one ton of bombs. Statistically, it was known that 50% of the planes were shot down, killing the pilots. Thus, the pilots knew that when they took on those missions they had a 50% survival rate.

Aiming to enhance the impact of the bombing sorties on the target islands, the Operations Research Officer offered the pilots two choices. The first choice was to go for a regular mission with a full load of fuel (i.e. two tons), and one ton of bombs. The second choice was to toss a coin. If the result was a "head," the pilot would be "off-the-hook," meaning he was free to go home and not participate in the mission. "Tails" would mean that the pilot was to go on with the mission but with only one ton of fuel, which was only good for a one-way trip, and two tons of bombs. In scenario 2, the pilots were to fly on a one-way, no-return trip. Clearly, the two choices are statistically equivalent. In each of the cases, the pilots had a 50% chance of survival, and yet the pilots consistently chose the first choice.

What would you have chosen?

Kenneth Arrow told me that as far as he could remember, the odds were actually very much *in favor* of the *one-way mission*. Statistically, the fraction of pilots that took the mission and returned was only 0.25%. Therefore, by accepting the Operations Research Officer's offer, the pilots would have *increased* their chances of survival from 0.25% to 0.5%.

Nevertheless, the pilots unanimously refused the offer of the Operations Research Officer. When asked during the individual interviews as to why they refused the offer, each one replied that *he felt that he was a much better pilot than average*, and therefore *he* would not be shot down.

Personal, biased assessment of the chances of survival abounds in everyday life. Every time someone drives a car at a high speed, the chances of a car crash resulting in death is high and yet we are somehow "programmed" to assess that these statistics apply to everyone else except us.

Long ago, I read in a preface of a book on the theory of traffic flow, dealing with the methods of designing bridges and highways, that it was easy to convince the reader that people, in general, drive at random speed, and make turns randomly at times, and at random places. It is impossible, however, to convince a specific individual that he or she is driving at random speed, making random turns at random times and in random places.

Our personal subjective assessments of probabilities are understandable. Also, the subjective feelings of *all* the pilots that they were *above* the average are understandable. This is different from the claim of the Highiq State University — that *all* the students had above-average grades.

Snack: The Virtuous Sheila on the Ferris Wheel

Sheila and Moshe, who recently married in an ultra-Orthodox Jewish ceremony, were walking leisurely in an amusement park one fine day during the Chanukah holiday. They could hear the shrill sound of children and adults screaming and were curious to find out where the screaming emanated from, until they finally saw an extremely huge Ferris wheel. Sheila became so revved up and excited, like a little girl she tugged on Moshe's sleeve and said: "Oh Moshe, I would love to try and ride that big wheel. It must be fun to be up there."

"You must be out of your mind! How could you even think about that? It's so windy up there and who knows what the strong wind can do, blow your skirt up like in Marilyn Monroe's infamous picture. If that happens your undergarment will be exposed for all the world to see!" Moshe said sternly.

That did not deter Sheila from tugging at his sleeve even harder and, with pleading eyes, saying: "Please Moshe, only once. I have never tried riding in this wheel before, please." Moshe was adamant. "No Sheila, I cannot have my wife's undergarments exposed to the public and be the talk of the town. More importantly, it's a grave sin!"

Before he could even continue to speak, Sheila was nowhere in sight. He looked around and called out her name but still there was no sign of her. "Oh well, she will find me."

From out of nowhere, he seemed to have heard his name being called. It was Sheila's voice but it was rather faint. Trying to trace where the voice was coming from, his ears led him up there to the Ferris wheel.

"Moshe, I am here. Oh, it's wonderful up here, come and join me!" Sheila declared ecstatically.

Moshe almost fainted at the sight of Sheila, with an ear-to-ear grin, waving at him from one of the Ferris wheel's passenger gondolas.

"Oy vey! That's an unforgivable sin. How could you have been so careless and unmindful? I told you, it's windy up there and your skirt might just be blown and expose your...."

Before he could finish his sentence, Sheila shouted back reassuringly, "No need to worry Moshe. *No one will see my undergarment.* I put it in my purse!" She waved her purse proudly, reassuring her husband that no one would see her undergarment.

Was Moshe himself so virtuous? How did he know of Marilyn Monroe's infamous picture?

The Maxwell Boltzmann distribution

And its application…

CHAPTER 5

Discover the Maxwell–Boltzmann Distribution

In this chapter, we shall carry out a new (and final) group of experiments. The structure of this chapter is similar to that of Chapter 4, and the experiments are almost the same. As in Chapter 4, we have a system of marbles distributed in cells. The experiment involves the "shaking" of the system, with exactly the same protocol as when we carried out the experiments in Chapters 3 and 4. However, unlike the experiments in Chapter 3, where we had no "string attached" to the marbles, and unlike the experiments of Chapter 4, where we had strings attached to the marbles, in this chapter we impose "cloth attached" to the marbles.

The reason for attaching cloth rather than string will become clear only in Chapter 7. For now, we will just describe the new condition imposed on the marble, do the experiments, and examine the evolution of the system. Or, if you prefer, think of the new game as being a little harder than the game in Chapter 4, which was harder than the game in Chapter 3.

Imagine that we attach a piece of square cloth to each marble. The area of the cloth is simply the square of the cell's level. Thus, the cloth attached to the marble in cell 1 (i.e. in level 0) has area 0. To the marbles in cell 2 (level 1), the attached cloth has a unit area. To cell 3, the area is $2^2 = 4$ and so on. In general, to the marble in cell k we attach a cloth with area $(k-1)^2$, as shown in Fig. 5.1.

The experiment we now carry out is essentially the same as in the previous chapters. We choose a marble at random and move it to a new, randomly chosen cell. We require that the total areas of all the cloths attached to the marbles in the system be conserved. This means that if we move a marble— say, from cell 2 to cell 3 — the cloth area changes from 1 to 4. Therefore, to conserve the total area of the cloth we must make another move — say, of three marbles from cell 2 to cell 1. An example of such a move is shown in Fig. 5.2.

Configuration: {4,3,2,1}
Number of Marbles: 10
Total cloth area: 1x0+3x1+2x4+1x9=20

Level: 0 1 2 3

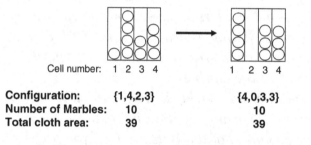

Level:	0	1	2	3
Area:	0	1	4	9

Fig. 5.1   The system with cloth attached to each marble.

Cell number: 1 2 3 4   1 2 3 4

Configuration:	{1,4,2,3}	{4,0,3,3}
Number of Marbles:	10	10
Total cloth area:	39	39

Fig. 5.2   A possible move from one configuration to another, conserving the total cloth area (here $TA = 39$).

Clearly, the experiment here is more difficult to simulate on a computer. Since we move marbles at random and select only those moves that conserve the total area, we shall be using a great number of moves which are not accepted as valid moves. In practice, using the same protocol for the simulation as in Chapter 4, it was found that on average, one out of a few thousand moves were accepted, depending on the number of marbles (NM) and the number of cells (NC). This fact imposes a limit on the size of the system that can be handled with the same program. Therefore, we shall be working only with small systems. Having experienced the systems in Chapters 3 and 4, we can easily extrapolate the results for larger systems.

A new feature of the system discussed in this chapter is that we allow cells with negative index, or levels; for instance, the levels $-3, -2, -1, 0, 1, 2, 3$. In all the experiments we shall actually carry out the simulation on the "one-sided" system with cells having positive numbers, (say, 1, 2, 3, 4) or levels (0, 1, 2, 3) and the corresponding cloth attached (0, 1, 4, 9), as in Fig. 5.2. However, our interest will be in the equivalent "two-sided" system with cells numbered $-3, -2, -1, 0, 1, 2, 3$, and the corresponding cloth areas (9, 4, 1, 0, 1, 4, 9). This is done only for convenience; the "one-sided" experiments are easier to carry out and take a shorter time to reach the equilibrium level. Note, however, that cells with negative index — say, –k — have a positive cloth attached area of k^2. Just to assuage your anxiety, allow me to give you a hint. The cell levels here represent the velocities of the particles in a one-dimensional system. The cloth area represents the kinetic energy of the particle. Clearly, the kinetic energy of a particle is the same if the particle moves forward or backward with velocity v or –v (more on this in Chapter 7).

As I have noted in Chapter 4, the experiments carried out in this chapter are more challenging than the ones in Chapter 3. The reader who is interested only in a rudimentary understanding of entropy and the Second Law can skip this chapter, as well as Chapter 4.

However, I urge you to read this chapter even superficially. You will learn not only about an important new distribution in physics, but also how this particular distribution arises from a random process with an additional constraint: here, the conservation of the total area, but in reality it is the conservation of the total kinetic energy of the particles. This is also discussed in Chapter 7.

5.1. Six Marbles in Three Cells

Remember the exercise in Section 3.1? In Section 4.1, we discussed the solution of that exercise along with the experiment of the same size but with strings attached. In the uniform case, there were seven possible configurations, and the one with the highest multiplicity was the uniform (2, 2, 2) distribution. In the present case, we have only two possible configurations that fulfill the conditions of total area equal to $TA = 6$. It is easy to calculate the multiplicity of these two configurations. These are 1 and 60 for the two configurations shown in Fig. 5.3. The corresponding values of the *SMI* for the configuration with the higher multiplicity is *SMI* = 1.49.

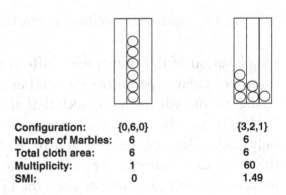

Configuration:	{0,6,0}	{3,2,1}
Number of Marbles:	6	6
Total cloth area:	6	6
Multiplicity:	1	60
SMI:	0	1.49

Fig. 5.3 All possible configurations for a system of (6;3).

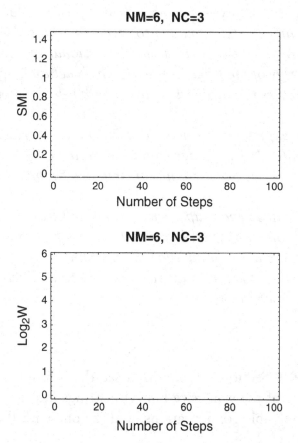

NM=6, NC=3

NM=6, NC=3

Fig. 5.4   The evolution of the system (6;3).

It is important to examine this relatively simple system before continuing to the larger system, where the calculations cannot be performed manually.

Figure 5.4 shows the evolution of this system. As can be seen from the figure, not much happens. The values of *SMI* oscillate between the two values 0 and 1.49, with the majority of visits, as expected, to the level with higher multiplicity. The same behavior is shown in the evolution of $\log_2 W$, which also attains only two values; 0 and 5.9, corresponding to W 1 and 60.

5.2. Nine Marbles in Varying Number of Cells

In this section, we fix the number of marbles at 9 and examine the evolution of the system with different number of cells (4, 6, 8). We choose for these particular experiments the total area of the attached cloth to be 9. The initial configuration will always be such that all the marbles are in cell 2 (or level 1).

Figure 5.5 shows the evolution of the system (9;4). As can be seen, there are only four values of the *SMI* and of *W* (the values of *W* are 1, 9, 252, 504). In this particular example, we observe only one visit to the initial configuration. In most of the steps, the system visits the upper levels which correspond to the larger multiplicities. The corresponding density profile is shown in Fig. 5.6a. In Fig. 5.6b, we also show the curve corresponding to the system with negative cell numbers. These

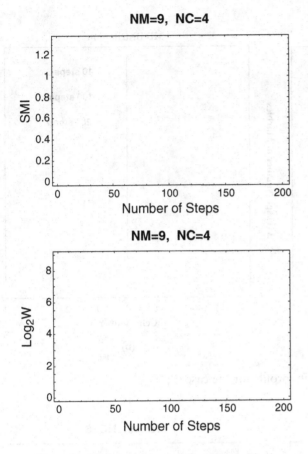

Fig. 5.5   The evolution of the system (9;4).

curves may be obtained either from a separate simulation or simply by taking the mirror image of Fig. 5.6a. In Fig. 5.6a, we show the initial as well as three other density distributions. Figure 5.6b was obtained by taking the mirror image of Fig. 5.6a. Note also that in Fig. 5.6a, the cell numbers are 1, 2, 3, 4, whereas in Fig. 5.6b, there are 7 cells denoted −3, −2, −1, 0, 1, 2, 3. If we want to obtain the symmetrical distribution, we must, in general, multiply the number of particles and change the total cloth area. For simplicity, we shall always do the "one-sided" experiment, then plot the mirror image, and renumber the cells. Figure 5.7 shows the evolution of the system, and Fig. 5.8 shows the density profiles for the case (9;6). As is clear from the figure, the additional cells were unoccupied.

In Fig. 5.8b, we show only the limiting density distribution. It is clear that this distribution is identical to the one in Fig. 5.6b.

This is a general result; once the tail of the density profile reaches the level below, say, 0.5, increasing the number of cells further will have no effect on the equilibrium density profile. The experiment with (9;8) took a longer time because many configurations were not accepted. However, the final distribution was identical to the one in Fig. 5.6.

The reason for this behavior is the same as the one discussed in Chapter 4. When we fix the total area (*TA*), there is a limit on how far the marbles can be spread.

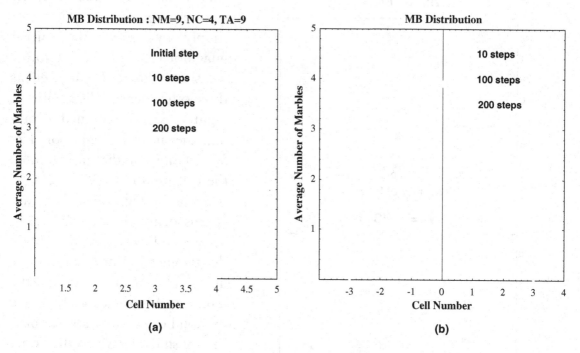

Fig. 5.6 The density profile for the case (9;4).

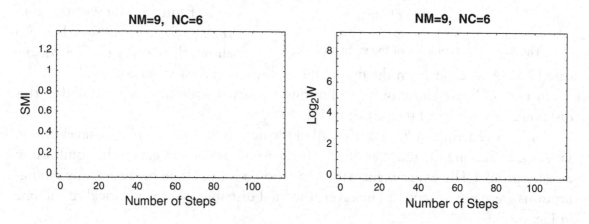

Fig. 5.7 The evolution of the case (9;6).

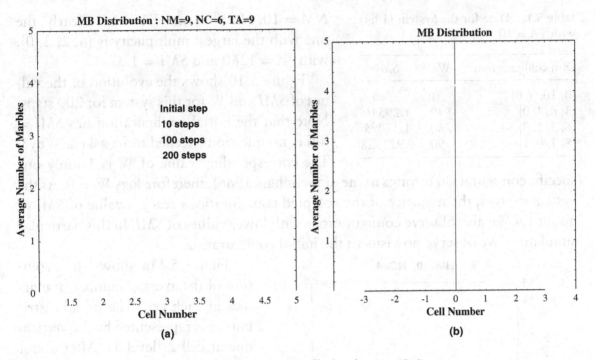

Fig. 5.8 The density profile for the case (9;6).

Therefore, no matter how large the number of cells is, the final equilibrium distribution of marbles in cells is determined by the fixed value of *TA*. Increasing the number of cells will only affect the simulation time since we will be performing many more steps on unacceptable configurations.

5.3. Chaning the Number of Marbles in a Fixed Number of Cells

Figure 5.9 shows the system with $NM = 10$ and $NC = 4$. In the initial configuration, all marbles are in cell 2 (or level 1), which means a total area of cloth $TA = 10$. Table 5.1 shows all possible configurations with

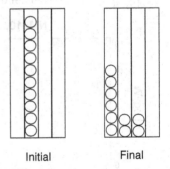

Initial   Final

Fig. 5.9 The initial and the final configuration for the system (10;4).

Table 5.1. Data for the System (10;4) with $TA = 10$

Dim-configuration	W	SMI
{0, 10, 0, 0}	1	0
{3, 6, 1, 0}	840	1.29546
{6, 2, 2, 0}	1260	1.37095
{8, 1, 0, 1}	90	0.921928

$NM = 10$, $NC = 4$, and $TA = 10$. Clearly, the one with the largest multiplicity is {6, 2, 2, 0}, with $W = 1260$ and $SMI = 1.37$.

Figure 5.10 shows the evolution of the values of SMI and W for this system for 200 steps. Note that the initial configuration has $SMI = 0$ (i.e. no questions needed to be asked. Why?) The corresponding value of W is 1 (only one specific configuration belongs to the dim-configuration), therefore $\log_2 W = 0$. As the system evolves, the majority of the accepted configurations reach a value of SMI of about 1.4 We also observe configurations with lower values of SMI. In this particular simulation, we observe no visits to the initial configuration.

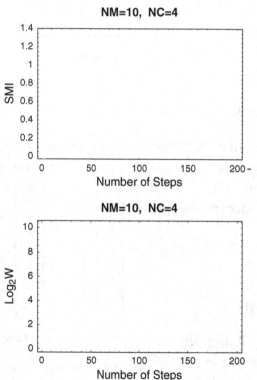

Fig. 5.10 The evolution of the system (10;4) in terms of SMI and W.

Figure 5.11a shows the evolution of the average number of marbles in each cell. The initial distribution is represented by the vertical line at cell 2 (level 1). After about 200 steps, the average number of marbles in each cell reaches an equilibrium value. Further steps do not have any effect on this equilibrium density profile. Fig. 5.11b shows the equivalent symmetrical distribution for a system with seven cells (−3, −2, −1, 0, 1, 2, 3), where the cell 0 is now level 0.

We next move on to the case of $NM = 20$ and $NC = 4$. Figure 5.12 shows the evolution of SMI and W for such systems. Note that here again the initial configuration is characterized by $SMI = 0$ and $W = 1$. The values of SMI and $\log_2 W$

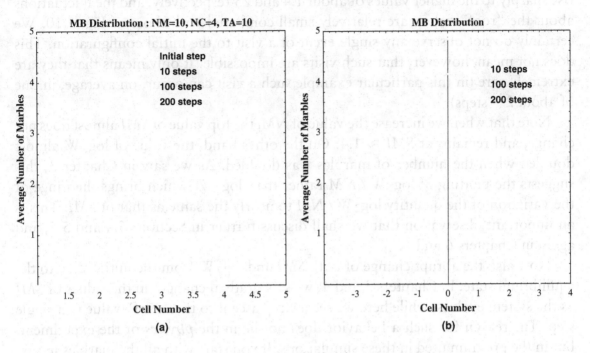

Fig. 5.11   The density profile for the system (10;4).

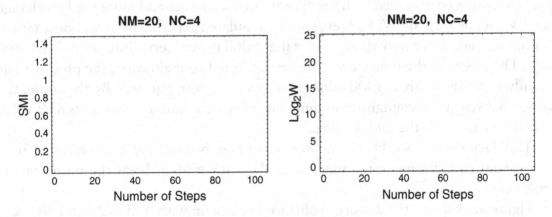

Fig. 5.12   The evolution of *SMI* and *W* for the system of (20;4).

rise sharply to the higher values of about 1.4 and 25 respectively, and the fluctuations about the "roof" values are relatively small compared with those of Fig. 5.10. We certainly do not observe any single event of a visit to the initial configuration. This does not mean, however, that such visits are impossible. It only means that they are extremely rare (in this particular example such a visit can occur, on average, in one of about 2^{23} steps).

Note that when we increase the value of NM, the top value of SMI almost does not change, and remains at $SMI \approx 1.4$. On the other hand, the value of $\log_2 W$ almost doubled when the number of marbles was doubled. As we saw in Chapter 3, this suggests the plotting of $\log_2 W / NM$ rather than $\log_2 W$, which brings the range of the variation of the quantity $\log_2 W / NM$ to nearly the same as that of SMI. This is an important observation that we shall discuss further in Sections 5.4 and 5.5, and again in Chapters 6 and 7.

Note also the abrupt change of both SMI and $\log_2 W$ from the initial state to the equilibrium state. In Chapters 3 and 4, we saw gradual changes in the values of SMI as the system evolves, while here we see a steep ascent to the "roof" value in a single step. The reason for such a behavior does not lie in the *physics* of the experiment, but in the program used in these simulations. If you start with all the marbles in say, level 1, and *physically* shake the system, then clearly the first move will be such that two or more marbles will move from level 1 to higher and lower levels (so that the total cloth area is conserved). This will result in a smoother and more gradual change in the values of SMI and W. However, in the simulated experiments, we choose moves at random and accept only those moves that fulfill the conservation of the fixed total area. Therefore, by the time we get the first accepted configuration, the program has already tested more than 1000 configurations that were rejected. By that time, the system has reached a configuration quite far from the initial one, hence its SMI value is also very far from the initial value.

This fact should not be of any concern to us because we are interested only in the final equilibrium state, and not in the path leading from the initial to the final state.

Figure 5.13 shows the density profile for the system with $NM = 20$ and $NC = 4$, and the corresponding figure with its mirror image. You can see that the overall form of the curve is nearly the same as in Fig. 5.11.

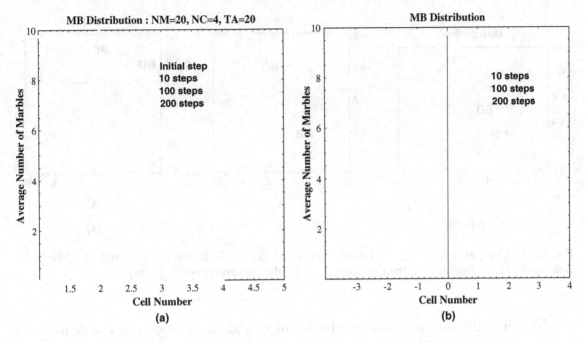

Fig. 5.13 The density profile for the system of (20;4).

We can do some experiments with larger numbers of marbles, but we shall not learn anything fundamentally new. However, with larger numbers of marbles, the number of steps wasted on unacceptable configurations becomes very large, and the number of accepted configurations becomes very small.

One can extrapolate from the two cases that we have studied here to larger numbers of marbles. We will surely find larger values of the multiplicities, but the equilibrium values of *SMI* will remain almost unchanged. We can also predict from the experiments carried out both in this section and in Chapters 3 and 4 that, as the number of marbles increases, the fluctuations around the equilibrium state of both the *SMI* and *W* become relatively small.

In all the experiments carried out in this and in the previous section, we have observed an equilibrium density distribution which starts at some maximal value at level zero, then decreases sharply at cells of higher levels (this is true if we look at either the one-sided or two-sided distributions). This is superficially similar to the distributions we observed in Chapter 4. However, because of the small number of

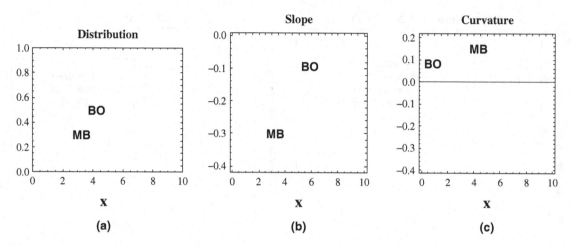

Fig. 5.14 Comparison of the analytical curves for (**a**) the Boltzmann (BO) and the Maxwell–Boltzmann (MB) distribution, (**b**) the slopes, and (**c**) the curvatures of the curves.

marbles, it is difficult to compare the forms of the distributions in the Boltzmann (BO) case and in the Maxwell–Boltzmann (MB) case.

One can show analytically that there are two important differences between the BO-distribution and the (one-sided) MB–distribution. The BO-distribution starts with a finite negative slope at level zero. The MB–distribution starts with zero slope at level zero (see Fig. 5.14a). Second, the curvature of the BO-case is always positive. On the other hand, in the MB–case the curvature changes sign; initially a negative, and then changes into a positive. Figure 5.14 shows the curves, the slopes, and the curvatures for the two distributions.

I have added these distributions here although they were not derived from the experiments we have carried out. The exact shapes of these curves are of no concern to us in our endeavor to "discover" the Second Law of Thermodynamics. However, it is helpful to know that had we done the simulations with very large numbers of marbles and large numbers of cells, we could obtain curves that are quite similar to the ones obtained analytically, as shown in Fig. 5.14. All the distributions in this chapter were derived from a small number of marbles and a small number of cells. This is the reason for having curves which look like polygons rather than normal smooth curves (see Note 1 and Fig. 5.22).

Before proceeding to the next two sections, I should mention that in this chapter we skip the simple processes such as expansion, mixing, and assimilation. These were important processes that we studied in Chapter 3. In Chapter 4, we saw that these processes do not provide any new results. The same is true for the systems discussed in this chapter, and therefore we do not need to do these experiments.

5.4. Dependence on the Total Cloth Area

In the previous two sections we examined the evolution of a few relatively small systems; that is, with small numbers of marbles and small numbers of cells. We need to stretch our imagination in order to extrapolate from these small systems to larger systems, which unfortunately cannot be handled even with the help of a computer.

However, from what we have learned so far, we can predict how a large system will behave. For any system with large number of marbles, the system will evolve towards equilibrium, at which both SMI and W will attain a maximum value, and the fluctuations about these levels will be relatively small. The density profile at the equilibrium state for the symmetrical system will look like the Normal distribution.[1]

In this section, we examine how the limiting equilibrium values of the SMI, W, and the limiting density profile change with the total area TA.

Before reading this and the next section, I urge you to read *and* understand Sections 4.5 and 4.6.

Figure 5.15 shows a series of density profiles for a system of $NM = 20$ marbles in $NC = 11$ cells,

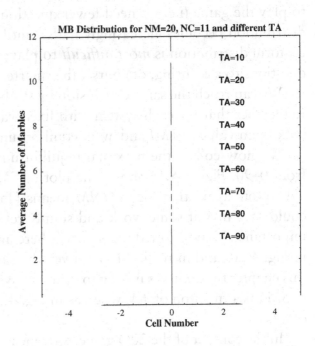

Fig. 5.15   The density profiles for different TA.

with different values of *TA*. We exclude the case $TA = 0$. This is the case where all the marbles are in level 0. In this case, there is only one configuration which is consistent with $TA = 0$, and that is the initial configuration. We can shake the system as vigorously as we wish and for as long as we wish, but if we keep $TA = 0$ fixed, nothing will happen to the system.[2]

Thus, we start with $TA = 10$ and increase its value by 10 units of area until we reach $TA = 90$. We see that as the *TA* increases, the curves become flatter and their widths become increasingly larger. Clearly, having more cloth area is interpreted here as having more possibilities for spreading the marbles over a wider range of cells. This is essentially the same phenomenon we observed in Section 4.5. At $TA = 200$, we obtain a flat curve and if we go beyond $TA = 200$, the curves become "inverted" curves. Again, this is similar to the "inverted" Boltzmann distributions that we discussed in Section 4.5. We shall not be interested in values of *TA* higher than 200.[3]

We recall that in terms of the 20Q game, a sharper distribution means it is easier to play the game (i.e. we need fewer questions on average). As we increase *TA*, the distribution becomes increasingly flatter until it becomes uniform. The game with a uniform distribution is *most difficult* to play; that is, we need the largest number of questions to ask (using, of course, the smartest strategy).

We can reach the same conclusion by studying the values of *SMI* as *TA* increases. Remember that for each system with fixed values of *NM*, *NC*, and *TA*, we reach a maximum value of *SMI* and *W* at equilibrium.

We now collect the maximal (equilibrium) values of *SMI* and *W* for each *TA* from 0–90. Figure 5.16 shows the plot of *SMI* and $(\log_2 W)/NM$ as a function of *TA*. (Note again that $\log_2 W/NM$ means $(\log_2 W)/NM$ and not $\log_2 (W/NM)$.) I could save myself some work and simply refer you to Fig. 4.20, but because of its importance I have plotted these curves here again. They look identical to the curves in Fig. 4.20, and in reality they convey the same information, but in the context of this chapter these curves belong to different systems. Note also that the plots here are of *SMI* as a function of *TA*, whereas in Fig. 4.20, the plots were drawn as a function of *TL*.

In the context of the 20Q game, *SMI* measures the "size" of the game; that is, the number of questions one needs to ask. Therefore, from Fig. 5.16 we can conclude

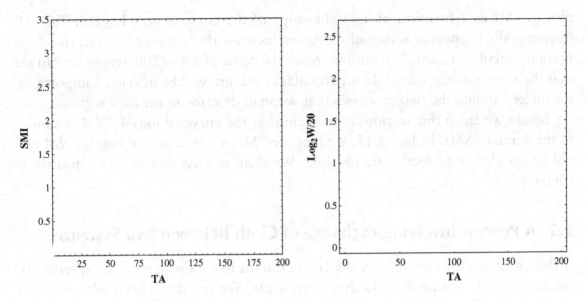

Fig. 5.16 The *SMI* and $\log_2 W/20$ as a function of *TA*.

that as we increase the value of *TA*, the "size" of the game becomes larger; that is, the game becomes increasingly more difficult to play. The curve starts at *SMI* = 0, which corresponds to the easiest game — no questions need to be asked, but as *TA* increases, more and more questions are needed, until we reach *TA* = 200, where we hit the most difficult game.[4]

As we have emphasized in Section 4.5, the monotonic increasing values of *SMI* with *TA* (or *TL*) is very important, but more important is the curvature of the

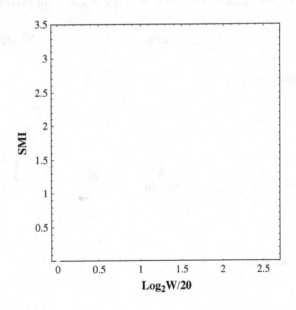

Fig. 5.17 The *SMI* as a function of $\log_2 W/NM$.

plot of *SMI* as a function of *TA*. The slope of the curve is very large at $TA = 0$ (theoretically it approaches infinity). As we increase the value of the *TA*, the slope becomes smaller and smaller, until we reach the value of $TA = 200$, where we can see that the slope becomes zero. This particular curvature will be of crucial importance for understanding the two processes that we shall describe in the next sections.

Before we leave this section, note again that the curve of $\log_2 W/NM$ is similar to the curve of *SMI*. In Fig. 5.17, we plot the *SMI* as a function of $\log_2 W/NM$ and obtain an almost perfectly straight line. We shall discuss this behavior further in Chapter 7.

5.5. A Process Involving Exchange of Cloth Between Two Systems

In this and in the next section, we shall carry out the two most important experiments. Actually, we do not need to do the experiments. We can draw from what we have already learned in Chapter 4, to infer on the outcome of these two experiments.

The first is quite simple. We start with two systems each with $NM = 20$ and $NC = 11$, as described in Fig. 5.18. The two systems differ in their *TA* value — one has $TA = 70$, the second has $TA = 10$.

The *initial* state of the two systems is such that they are already at their equilibrium states. That means that we have already performed the equilibration process for each of the two systems. Therefore, we know the initial value of the *SMI* and *W* for each of these systems at equilibrium. These values may be read from Fig. 5.16.

We now "transfer" 20 units of cloth from the

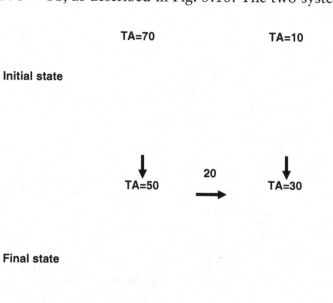

Fig. 5.18   A process of transferring of 20 units of cloth.

system with $TA = 70$ to the system with $TA = 10$. This simply means that we change the values of TA for the two systems, one from 70 to 50, and the second from 10 to 30. The two systems are now allowed to reach a new equilibrium state. Why new? Simply because we know that a system with $TA = 50$ will have a different equilibrium state, with different values of SMI and W, than the system with $TA = 70$. The same applies to the system that was initially at $TA = 10$, and now has the value $TA = 30$. We also know the values of SMI for the new systems; these can be read from Fig. 5.16.

Because of the particular curvature of SMI as a function of TA, the decrement of SMI of the first system is smaller than the *increment* of the SMI of the second system. The net change in SMI is thus positive. We can conclude that transferring of a *small* amount of cloth from the system with the high value of TA to a system with the low TA value will always result in a *net increase* in SMI. The detailed argument is exactly the same as in the process of transferring string from one system to another (see Fig. 4.23 in Chapter 4). The analysis made in Chapter 4 using Figs. 4.23 and 4.26 applies to this case as well, and will therefore not be repeated.

Next, we examine the change in the shape of the density distribution in this process. In Fig. 5.15, we drew a series of density distributions for different values of TA. As can be seen from both Figs. 5.18 and 5.19, the distribution of the $TA = 10$ system became flatter, while the distribution of the $TA = 70$ system became sharper. This result is of the greatest importance in understanding how the Second Law "operates" in a process of heat transfer. This topic will be discussed in Chapter 7.

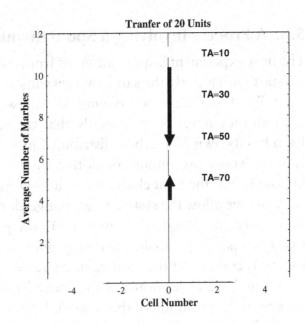

Fig. 5.19 The change in the density distribution in the process of transferring 20 units of cloth.

In the language of the 20Q game, we can rephrase the same conclusion as follows: Initially, we had two systems, one (with $TA = 70$) being more difficult to play than the other (with $TA = 10$). After transferring 20 units of cloth, the first game became easier to play, but the second became more difficult. However, when evolved to their new equilibrium states, the net change is such that the combined system will always be more difficult to play than the initial combined system.

Before we continue to the next (and final) experiment, it is important to understand what *we have done to* the system, and what the system *has done* in response. We started with two systems, each in its own equilibrium state, and we then transferred 20 units of cloth from one system to the other. If you do not like the word "transferred," simply imagine that we *increased* the value of TA of one system from $TA = 10$ to $TA = 30$, and *decreased* the value of TA of the second system from $TA = 70$ to $TA = 50$. After doing this, we allow each of the systems to reach a new equilibrium state. The results of this transfer process are shown in Fig. 5.18.

5.6. A Process Involving a Spontaneous Transfer of Cloth

The next experiment is quite different from the one described in Section 5.5. Initially, we start with exactly the same two systems as before (Fig. 5.20). Both are at equilibrium. We do not transfer *anything*. We allow the two systems to *share* the total area of cloth they have. Note carefully that the two systems do not exchange marbles. Each has its own 20 marbles distributed in 11 cells. What we allow the system to do is to *exchange* any amount of cloth *they wish* to exchange. The only restriction we impose is that the *total* cloth area will be conserved at $TA = 70 + 10 = 80$.

Once we allow the system to exchange cloth, we start to "shake" the two systems simultaneously. We do not intervene in the process of sharing of cloth between the two systems, we just shake the two systems so that they reach a new equilibrium state together, conserving the total value of $TA = 70 + 10 = 80$.

If you have never heard of the Second Law you will have an eerie feeling while observing the evolution of this system. Remember that the two systems were initially at equilibrium, each having a maximum value of SMI and W. When you start to "shake" the combined two systems simultaneously, you will observe that cloth will "fly" *spontaneously* from the system with high value of TA to the system with low

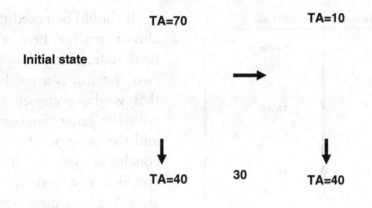

Fig. 5.20 The spontaneous process of transfer of cloth between two systems.

value of *TA*. This voluntary exchange of cloth will continue until the two systems reach a new equilibrium state. At that state the value of *TA* of each system will be *TA* = 40; that is, the high *TA* system has voluntarily transferred 30 units of cloth to the low *TA* system, which voluntarily accepted that donation. The eerie thing about this process is that we *did not* intervene in the transfer of cloth between the two systems, nor did we determine how much cloth would be exchanged between the two systems.

Rightfully, you may ask how I knew what was going to happen to the system. A quick answer is: Try it! Simulate the experiment on your computer and see what happens. I shall answer this question in more detail in Chapters 6 and 7. It is concerned with the shape of the curve of *W* (or $\log_2 W$) as a function of *TA*. At this moment, I urge you to accept my word and concentrate on *what* has happened rather than on the question of *why* it happened.

Figure 5.21 shows the density distribution for the two systems in the initial state, and in the final state. This is similar to Fig. 5.19, except that the final states of the two systems are the same.

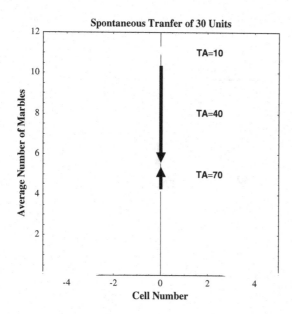

Fig. 5.21   The change in the density distribution in the process of spontaneous transfer of cloth between the two systems.

It should be noted that the conclusion reached here regarding the final state, being the same for the two systems, is a result of the fact that we have chosen two systems with the same number of marbles and the same number of cells. The conclusion will be different if we combine two systems of different sizes. We have discussed the case of two different systems in Section 4.8. Exactly the same analysis applies to this case.

5.7. What Have We Learned in this Chapter?

In Chapter 3, we saw that all the equilibrium density distributions were uniform. In Chapter 4, the density distributions were exponential; that is, the Boltzmann-type distributions. These distributions may be obtained by maximizing the *SMI* subject to the condition that some average quantity is given. Similarly, the Normal distributions[1] may be obtained by maximizing the *SMI* subject to the condition that the *variance* or the standard deviation is given. This theorem was proven by Shannon. There is a fundamental difference between the exponential distribution and the normal distribution. Nevertheless, one can obtain one from the other by transformation of variables. This is discussed further in Chapter 7.

As in Chapters 3 and 4, we noticed that when we increase the number of marbles, the fluctuations of both *SMI* and *W* about the equilibrium state become relatively small.

As in Chapter 4, we found that when we increase the number of cells, the density distribution does not change (provided that the number of cells is large enough so that the average density at the most distant cells is nearly zero).

Finally, in Chapters 4 and 5 we observed a spontaneous process conserving one thing (total string, total cloth) but not conserving another (*SMI, W*). We shall see in Chapter 7 that these observations are echoes of the Second Law.

Snack: Klutz and Botch

Who is the Schlemiel and who is the Schlimazel?

Do you know the difference between a schlemiel and a schlimazel?

If you do not know what these words mean, you can look them up in an online dictionary. These are very useful concepts, in particular for those who associate entropy with disorder.

These two concepts can be defined in one sentence:

A *schlemiel* is an unfortunate man who when carrying a hot bowl of soup, will always end up dropping the piping hot soup — on the lap of the unfortunate *schlimazel*.

Snack: The Importance of Units

Entropy was originally defined as a quantity having units of energy and temperature. Entropy is now interpreted as a measure of information — in units of bits.

These seemingly "conflicting" units reminds me of a story my brother told me when he used to work as a teacher in a special education school. Every week he and his fellow teachers would meet to discuss the progress of the students, as well as any other issues which needed to be addressed and brought to the attention of their headmaster, who was a psychologist by profession.

In one of those meetings, my brother raised the issue of one of his students, Max Lowiq (not his real name), who was lagging behind, who never seemed to understand anything, had very poor grades in his tests, never did his homework, was inattentive in class, and hardly ever spoke. My brother told the headmaster that he felt as if all his efforts were in vain.

Upon hearing this, the headmaster took out the file of Max Lowiq and seemed to be puzzled at what he read. Scratching his head, the headmaster said: "But I don't understand. I have a report here that Max has an IQ of 130. That is certainly a very high IQ."

Yes," my brother said, "but that IQ is in Fahrenheit!"

CHAPTER 6

Entropy and the Second Law in the World of Marbles

I would like to believe that you have reached this chapter in the book by not skipping all the tedious and tiring experiments in Chapters 3–5. However, if you read Chapter 3 carefully but skipped Chapters 4 and 5, you can still read this and the next chapter and obtain a basic feeling for what entropy is and why it behaves in such a peculiar way. On the other hand, if you did not even read Chapter 3, then there is no point in you reading this chapter. In this chapter, we shall examine, analyze, and try to reach some general conclusions based on the data we have collected in Chapters 3–5. This is the Normal agenda of a research scientist; first, to collect experimental data. This part can be long and tedious. The next stage is to analyze the data, search for some regularities, common trends or patterns, then try to interpret these, and perhaps discover a new law.

In this chapter, we shall analyze the data that we collected in Chapters 3–5. We shall eventually reach a general conclusion which we may refer to as an echo of the Second Law reflected from the world of marbles. In Chapter 7, we shall translate the content of this chapter from the language of marbles in cells into the language of real particles in real boxes. In both of these chapters, we shall refrain from using the tools of mathematics. Some mathematical "supportive" notes will be provided in Chapter 8. If you are allergic to the language of mathematics, you can skip Chapter 8 altogether and trust the validity of the conclusions reached in this and in the next chapter.

6.1. Summary of the Experiments with a Single System

In this section, we use the following two short-hand notations: Instead of referring to a game with, say, $NM = 9$ marbles in $NC = 3$ cells, we shall simply say the

(9;3) game. Instead of repeating the long description of the procedure of starting from an initial state, picking up a marble at random and moving it to a randomly chosen new cell and so on, we shall simply say that we *shake* the system. Finally, instead of referring to the *type* of constraints we had imposed in the experiments in Chapters 3–5, we shall simply refer to the *U*-case for the experiments where no constraints were imposed, resulting in the *uniform* distribution; the *BO*-case, for the case discussed in Chapter 4, resulting in the Boltzmann distribution; and the *MB*-case, for the case discussed in Chapter 5, resulting in the Normal distribution. However, since we are interested in the Normal distribution of *velocities*, we shall refer to this distribution as the Maxwell–Boltzmann distribution.

6.1.1. *The Uniform Distribution: Shannon's First Theorem*

In all the experiments carried out in Chapter 3, we shook the system until it reached an equilibrium state. The equilibrium state was characterized by having nearly constant values of all the parameters that we chose to follow and record during the evolution of the system. Specifically, the equilibrium state was characterized by maximum values both of the multiplicity, W, the probability, PR, and the Shannon measure of information, SMI. We have also observed that when starting with any arbitrarily chosen initial state, the system will evolve in such a way that the equilibrium density distribution (i.e. average number of marbles in each cell) will be *uniform*.

What we have observed is essentially an "experimental" verification of Shannon's first theorem.[1] For our games, this theorem states that of all possible distributions of marbles in cells, the uniform distribution is the one that maximizes the value of SMI. When the number of marbles is small, we observe fluctuations about the uniform distribution. However, when the number of marbles is very large, the relative fluctuations are so small that we cannot observe any fluctuations about the equilibrium distribution. The system will always spread its marbles over all the cells uniformly. For a given number of marbles (NM), the larger the number of cells (NC), the smaller the average number of marbles per cell will be. The average density at equilibrium will be NM/NC.

Thus, at the equilibrium state, the density profile remains uniform and unchanged. We also observed that when the number of marbles was large, the values of W

became very large. Therefore, for reasons of convenience only, we decided to follow $\log_2 W$ instead of W itself. For even larger numbers of marbles, we noticed that $\log_2 W$ roughly doubles when the number of marbles is doubled. This finding suggests that perhaps it will be even more convenient to plot $\log_2 W/NM$; that is, the quantity $\log_2 W$ per marble, instead of $\log_2 W$ (note again that $\log_2 W/NM$ means $(\log_2 W)/NM$). By doing that we discovered that the evolution $\log_2 W/NM$ was almost identical to the evolution of *SMI*. Moreover, the maximum value of *SMI*, reached after many steps (i.e. when the curve leveled off and remained nearly constant) was almost the same as the equilibrium value of $\log_2 W/NM$.

It should be emphasized again that plotting of $\log_2 W/NM$ was initially done merely for reasons of convenience. For very large numbers it is far more convenient to use the logarithm of a number rather than the number itself. However, the plotting of $\log_2 W/NM$ led us serendipitously to an important discovery: The equilibrium values of $\log_2 W/NM$ were roughly the same as those of *SMI*. This is a new discovery, because the quantities W and *SMI* seem to have nothing in common. W is the number of specific arrangements of marbles in cells that belong to a given dim-arrangement. On the other hand, *SMI* is the average number of questions we need to ask in the smartest strategy to find out where a specific marble is "hidden". Is there any deep and unexpected connection between these two quantities?

Figure 6.1 shows a plot of $NM \times SMI$ versus $\log_2 W$ for different ranges of values of NM. We chose $NM \times SMI$ rather than *SMI* for the following reason. *SMI*, as you remember, was a measure of the "size" of the game when we had to find the cell in which *one* marble was hidden. If we are asked to do the same for each marble, then we have to ask $NM \times SMI$ questions. We shall tentatively refer to this quantity as the *M-entropy* (for marble-entropy). In the next chapter, the *M-entropy* will be translated into the entropy of the system.

Figure 6.1 shows that for a small number of marbles, the values of $NM \times SMI$ and $\log_2 W$ are not quite the same. They both increase as NM increases, but they are not identical. For convenience, we show in the figure the diagonal line, that is, the line for which the values of the ordinate and the abscissa are equal. As we increase the number of marbles, the curve of $NM \times SMI$ versus $\log_2 W$ becomes increasingly closer to the diagonal line on the scale of this figure. For $NM \approx 1000$, the curve is almost identical to the diagonal line.[2]

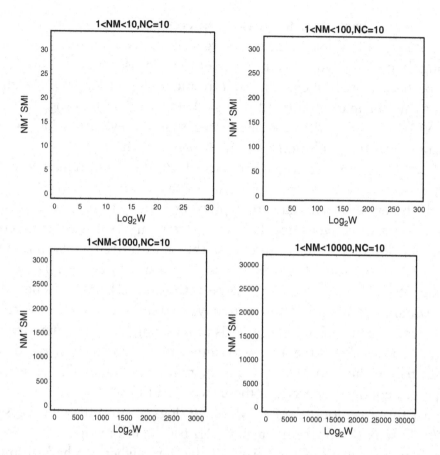

Fig. 6.1 $NM \times SMI$ as a function of $\log_2 W$ for different ranges of NM.

What we have found is quite remarkable — the equilibrium value of the *SMI* is almost the same as the equilibrium value of $\log_2 W/NM$. This remarkable result is at the heart of the Boltzmann formulation of the entropy. We shall discuss this formulation further in the next chapter. For the moment we can make a statement on the analogue of the Second Law for the marble game.

For each game of many marbles in cells, and for any initial arrangement, the value of the *SMI* will always increase when the system is shaken for a long time (many steps), until it reaches some maximal value then remains constant. Further shaking does not have any noticeable effect on the value of *SMI*. Furthermore, the value of *SMI* at equilibrium is nearly equal to the value of $\log_2 W/NM$.

In the language of the 20Q game, we can say that if we start with any game of marbles in cells, shaking the game makes the game more and more difficult to play (more questions), until the system reaches an equilibrium state. In that state the game is the most difficult to play (i.e. the distribution is uniform; remember NM and NC are fixed for the game). At equilibrium, we have the relationship $NM \times SMI \approx \log_2 W$.

This equality establishes the meaning of *what* is the SMI of the system in terms of the multiplicity W (at equilibrium). However, this relationship has more to it. It also harbors the answer to the *why* question. In all the experiments carried out in Chapter 3, we observed the steady ascent of the quantity SMI until reaching equilibrium. But why does the system always changes in that direction so as to make our game more and more difficult?

The answer to this question is probabilistic, and the probability of that event to occur is determined by the values of W.[3]

In our game, W is the number of specific arrangements belonging to a dim-state. While we shake the system, the dim-state changes at each step. The larger the value of W, the larger the probability of that dim-state. At equilibrium, the value of W becomes very large compared with all other values of W of the system (as we observed in the plots of W as a function of the number of steps). Therefore, the probabilities of these states become very large, too. This means that the system will spend more time in states having higher probabilities than in states having lower probabilities. When the number of marbles is very large, the equilibrium values of W become very large, too, and the probability of finding the system in the equilibrium state is nearly one.[4]

We can thus conclude that as the system (of many marbles) reaches an equilibrium state, its SMI value is maximum. Also, its W value is maximum. Both can be used to characterize the equilibrium state. However, W is the quantity related to the probability of the state, and therefore harbors the answer to the question of *why* the system proceeds towards the equilibrium state.

6.1.2. *The Boltzmann Distribution: Shannon's Second Theorem*

In Chapter 4, we carried out a series of experiments with marbles in cells. Strings were attached to each of the marbles. The length of the string attached to cell number k

(from the left cell numbered 1) was $k - 1$ (e.g. marbles in cell 1 having string length 0). We shook the system, while requiring that the total string length be *conserved*. That restricts the possible arrangements that the system can attain. Again, we observed that the system evolved towards an equilibrium state which was characterized by maximum values of W and *SMI*. However, the equilibrium density distribution was different in this case.

In all the games of Chapter 4, we obtained the Boltzmann (or the exponential) distribution of marbles, such that the density of marbles in the first cell is the largest, and the average density falls sharply as the distance of the cell from *ground zero* increases. Again, we observe fluctuations about the equilibrium state. The larger the number of marbles, the smaller the relative fluctuations about the equilibrium state. At equilibrium, the density profile becomes smooth and has the characteristic Boltzmann distribution. What we have observed is essentially Shannon's second theorem.[5] It states that of all possible distributions of marbles in cells, with the additional constraint that the total string length be constant, the exponential, or Boltzmann's distribution is the distribution that maximizes the *SMI*.

When we increase the number of cells but keep the number of marbles fixed, we might initially observe some changes in the equilibrium density distribution. However, beyond certain numbers of cells the density distribution remains unchanged, independently of the number of cells. Any additional cells would have no effect on the distribution of the marbles. The marbles will always be more densely packed in level 0, less dense in level 1, and so forth. This is in sharp contrast to the behavior of the uniform case.

The new feature of the experiments carried out in Chapter 4 was the dependence on the total string length. We found that the larger the string length, the flatter the density distribution is, up to a point when the density distribution becomes uniform.

In the language of the 20Q game, we can interpret this result as follows. Suppose we play the (9;3) game, and the total string length is 0, then all the marbles must be in level 0. No matter how much you shake the system, it will remain at the same state. How many questions do you need to ask to play this game? None. You know where all the marbles are, therefore you do not need to ask any questions. Thus, this game is the easiest game to play.

Next, suppose that $TL = 1$. This means that in the initial state 8 marbles are in level 0 and 1 marble is in level 1. This game is a little more difficult to play, but still you can win by asking, on average, about one question only.

As we increase the string length until we get to total length $TL = 9$, the game will be increasingly more difficult to play. It will be the hardest game to play when $TL = 9$. At this string length, the distribution becomes uniform.

In Section 4.4, we also examined the dependence of SMI and W on the string length TL. We found that both SMI and W increase monotonically with TL, up to a certain value of TL where the plots leveled off.

We can now make another statement of the Second Law applied to marbles with strings attached:

> Starting from any initial arrangement of the marbles in cells, after a long shaking period the system will reach an equilibrium state. This equilibrium state is characterized by maximum values of SMI and of W. The density distribution will always have the shape of the Boltzmann (or exponential) distribution. This means that as the system evolves towards equilibrium, the game becomes increasingly harder to play until we reach the equilibrium state where the game is the hardest (for the fixed value of TL).

For different values of TL, the maximal equilibrium value of SMI also increases with TL up to a point (see Fig. 4.23). The larger the value of TL, the more difficult is the equilibrium game.

Regarding the question of "why," we can say again that the system will evolve towards the state of higher probability. The probability is related to the multiplicity W, which we have observed reaching a maximal value at equilibrium. The exact connection between the probability and W is somewhat different in this case from the case discussed in Section 6.2.1 (see further discussion in Note 6).

6.1.3. *The Maxwell–Boltzmann Distribution: Shannon's Third Theorem*

The conclusions regarding the MB case are similar to the ones we arrived at in Subsection 6.1.2. In fact, one can formally derive all the conclusions for this case from the conclusions arrived at in 6.1.2, simply by means of transformation of variables.[7] Therefore, we shall be brief in this case.

As in the Boltzmann case, shaking the cloth-attached system will bring it to a state of equilibrium after many steps. Again, this state of equilibrium is characterized by a maximum value of *SMI* and *W*. The equilibrium density profile will have the typical form of the Normal distribution, which in the case of real particles is referred to as the Maxwell–Boltzmann distribution (see Note 8).

As in the Boltzmann case, when the number of marbles increases we shall observe relatively smaller fluctuations about the equilibrium state. On the other hand, if we increase the number of cells, initially we shall see some changes in the density profile, but further increase in the number of cells will have no effect on the density distribution. The marbles will concentrate at level 0, and the density drops sharply as we get further away from level 0, in whichever direction we move.

The dependence of the density profile and the values of *SMI* and *W* on the total cloth-area has been discussed in Section 5.4. We can now make a statement regarding the Second Law for marbles-in-cells having a fixed cloth-area.

> Starting from any initial arrangement of marbles in cells, shaking the system will lead to an equilibrium state. The equilibrium state is characterized by maximal values of *SMI* and *W*. The density profile of the equilibrium state will be the Maxwell–Boltzmann distribution.

For different values of the total area *TA*, the larger the *TA* the flatter the distribution is, up to a point at which the distribution becomes uniform.

A plot of the equilibrium values of *SMI* and *W* for different *TA* values is almost the same as the plots for the Boltzmann case. Also, the answer to the question of why the system proceeds to the equilibrium state is the same as provided in Subsection 6.1.2.

6.2. Summary and Conclusions Regarding the Processes Involving Two Systems

In the previous section, we summarized the results of our experiments on a single system. We now move on to the more interesting and more important processes involving two (or more) systems which are brought into "contact". The "contact" has different meanings for the different cases. Therefore, we shall discuss each case separately. This section is the most important one for understanding the Second Law in the world of marbles.

6.2.1. *Evolution of a Pair of Uniformly Distributed Games*

In Section 3.3, we examined several processes involving two systems that were first brought to their equilibrium state separately, and then the two systems were combined and allowed to reach a new equilibrium state. Let us examine each of these processes separately.

The expansion process

Two systems, one with NM marbles in NC cells and the second empty system having NC cells, are first equilibrated. The two systems are then combined to form one system of NM marbles in $2NC$ cells. Next, the combined system is shaken together allowing exchange of *marbles* between the two systems. Since the initial state was an equilibrium state, the density profile of the first system at equilibrium was NM/NC per cell. The density profile of the second cell was initially 0 (i.e. $0/NC$ per cell).

After combining and equilibrating the system, we found that the new equilibrated system had a uniform distribution with $NM/2NC$ marbles per cell; that is, half of the original density of marbles in the first single system. Thus, combining the two systems into one results in the *expansion* of the marbles that originally were confined in NC cells to the entire new system of the $2NC$ cells. We also recorded the *SMI* values of the entire system before and after the combination of the two single systems. Whatever the numbers NM and NC were, the value of *SMI* before the expansion occurred was $\log_2 NC$ (we had NC possibilities of finding a specific marble, therefore the average number of questions, or the *SMI*, is $\log_2 NC$). After the expansion there are $2NC$ cells, hence the *SMI* value is now $\log_2 2NC$. The net change in the *SMI* in the expansion process is $\log_2 2NC - \log_2 NC = \log_2 2 = 1$.

We can immediately formulate the Second Law regarding the expansion of marbles.

> In any expansion process the marbles will spread from their original cells to occupy the total number of cells of the combined system, and the final density distribution will be a new uniform distribution. The *SMI* of the system will always increase in this process.

Exactly the same conclusion can be drawn when both of the two systems are not empty — say, NM_1 in NC_1 cells and NM_2 in NC_2 cells. The final density profile will be, on average, $(NM_1 + NM_2)/(NC_1 + NC_2)$ marbles per cell.

Why has the system evolved to the new equilibrium system? Simply because the value of W has increased and W is related to the probability of the state. The larger the value of W (for the combined system), the larger the probability of the new equilibrium state of the combined system.

Pure mixing

As we have observed in Section 3.3, when we combine NM blue marbles in NC cells and NM yellow marbles in NC cells into a new system of NM blue and NM yellow in NC cells, the density profile of each of the colors of marbles will not change. The SMI values of the entire system will not change either. The new combined game is as difficult (or as easy) to play as the initial combined game.

Pure assimilation

Here, we combine NM marbles in NC cells with NM marbles of the same kind (color) in NC cells to form a combined system of $2NM$ marbles in NC cells. In this process, clearly the equilibrium density of distribution of the combined system will be doubled (i.e. $2NM/NC$) (see Fig. 3.28). The SMI value in this process will be reduced from $\log_2 NC + \log_2 NC$ to $\log_2 NC$.

Assimilation and expansion

In this process, we combine two systems of NM marbles in NC cells into a system of $2NM$ marbles in $2NC$ cells. Clearly, the density profile and the SMI value do not change in this process. Additionally, the SMI value of the combined system does not change.

Mixing and expansion

Combining NM blue marbles in NC cells with NM yellow marbles in NC cells into one system of NM blue and NM yellow marbles in $2NC$ cells is completely equivalent to two processes of expansion. The blue marbles expand from NC to $2NC$, and the yellow marbles expand from NC to $2NC$. Therefore, the change in SMI is simply twice the change in SMI in the expansion process, namely $2 \log_2 2 = 2$.

6.2.2. *Evolution of a Pair of Boltzmann-Type Games*

In Section 4.5, we conducted two experiments involving exchange of strings between two systems. In the first, we started with two systems of the *BO*-case at equilibrium. Then we transferred a specific amount of string from one system (with the higher *TL* value) to the other (with the lower *TL* value). After this transfer, the two systems were disconnected (i.e. no further *interaction*) between the two systems. We then let the two systems approach a new equilibrium state.

What we have found could partially be deduced from what we have learned in Section 6.1.2. We know that the equilibrium values of both *SMI* and *W* increase (up to a point) with the increase in *TL*. Therefore, we can expect that the system which donated the string will end up with lower values of *SMI* and *W*, whereas the system that received the string will end up having higher values of *SMI* and *W*. In the language of the 20Q game, we can say that the first system (the donor) became *easier*, but the second system (the acceptor) became *harder* to play.

What we have also found, which was unexpected, is that the *net* amount of change in both *SMI* and *W* was *positive*. This means that although one game became easier and the second one became harder, the two games combined became *harder* to play. This behavior followed from the particular curvature of the plot of *SMI* (and *W*) as a function of *TL*. We can now make another statement of the Second Law pertaining to this process.

> When a small amount of string is transferred from the game with higher *TL* to the game with lower *TL*, and the system is allowed to reach a new equilibrium state, the pair of games always becomes harder to play.

We added the quantifier "small amount" of string because we do not want to transfer too much of the string; for instance, if we transfer all the string length of the first system to the second, the Second Law will not apply. This can be seen from Fig. 4.23, and it will become clearer from the discussion below.

The second experiment was similar to the first, but was different in a very fundamental way. Again, we started with two systems (one with higher *TL* value — say TL_1, and one with lower *TL* values — say TL_2) and let them reach equilibrium states separately. In this equilibration process, each system had kept its own *TL* fixed. Each reached an equilibrium state with its own maximal values of *SMI* and *W*.

We next connected the two systems in such a way that string can *flow* spontaneously from one system to the other. In other words, we let the pair of system reach a new equilibrium state (by shaking them simultaneously), and requiring that the *total* string length $TL_1 + TL_2$ be kept constant.

What we have found is anything but expected. We saw that string *flew* spontaneously from the system with higher value of TL to the system of lower value of TL. At the final equilibrium state, the total values of SMI were *larger* than the sum of the initial values of SMI of the two systems.

Before we try to understand why this process has occurred spontaneously, let us make another formulation of the Second Law for the games of the BO-case.

> When two games of equal sizes (i.e. the same number of marbles and the same number of cells), each at equilibrium, having different values of TL, are brought to a new equilibrium state under the constraint that the total string length is conserved, string will always flow from the system having higher value of TL to the system having lower value of TL. The value of the SMI of the combined system in the final equilibrium state will be larger than the value of the SMI of the two systems in the initial equilibrium state.

One quick caveat: We chose two *equal games* for convenience. With two equal games we can *read* the values of SMI from the *same* plot of SMI as a function of TL (Fig. 4.23). If the two systems have different sizes, then we need to *read* the values of the SMI from two different plots; one for the first system, one for the second. An example is provided in Fig. 4.27. Here, for simplicity we assume that the two systems are equal, except for having two different values of TL.

Now for the question of *why*. Why will the system change in such a way that string will flow from the higher valued TL to the lower value of TL? The answer, as usual, is probabilistic. The new equilibrium state of the combined system has higher probability than the initial state.[9] This can be read from the plots of $\log_2 W$ as a function of TL (e.g. Fig. 4.26). Note again that we are dealing with two systems of equal sizes (i.e. NM and NC are the same). In general, we need to look at the plots of $\log_2 W$ for the two systems, as we have discussed in Section 4.8.

As we have seen in Section 4.7, $\log_2 W$ of one system has increased, and of the second has decreased, but the sum of the two values of $\log_2 W$ has increased. Now we use the property of the logarithm; that the logarithm of the *product* of the two numbers is the *sum* of the logarithm of the two numbers. Thus, $\log_2 W_1 + \log_2 W_2$ is the same as $\log_2 W_1 W_2$. The product of $W_1 W_2$ is the number of specific states of the combined system. It is this quantity that is related to the probability of the state of the combined system. In other words, the probability of the new equilibrium state of the combined system will be larger than the probability of the combined system in the initial state.

This reasoning answers the question of why the system will evolve in such a way that string will flow from the higher *TL*-value system to the lower *TL*-value system. This flow will continue until we reach a maximum value of the probability of the combined system. The maximum is reached when the two systems have equal value of *TL*, which in this case is $(TL_1 + TL_2)/2$. Any further flow of string will lower the probability (see Fig. 4.29).

It should be stressed again that we are discussing two equal-sized systems. For two systems of unequal size, the argument is more lengthy and cumbersome (see Section 4.8).

6.2.3. *Evolution of a Pair of Maxwell–Boltzmann Type Games*

The third case is the Maxwell–Boltzmann (*MB*) type of system. The analysis and conclusions of the processes involving the transferring of cloth between the two systems is essentially the same as in the *BO*-case. Therefore, we shall be very brief in discussing this case.

Regarding the transfer of a small quantity of cloth from the system with a high value of *TA* to the lower value of *TA*, the relevant formulation of the Second Law is exactly the same as in Section 3.2.2; we have only to replace the *TL* with *TA*. The Second Law applied to the *MB*-case is therefore:

> When a small amount of cloth is transferred from the game with higher value of TA to a game with lower value of TA, and the system is allowed to reach a new equilibrium state, the pair of games will always evolve towards a harder game.

Again, we comment that this is true only for a transfer of a small amount of cloth. How much is a small amount? Not more than the amount that will be transferred spontaneously, as in the next experiment.

The second experiment we carried out in Section 5.6 was the *spontaneous* transfer of cloth from one system to another. Again, all the discussion of Section 6.2.2 may be carried over to this case. The only difference is replacing string by cloth, and *TL* by *TA*. The conclusion in this case is also the same as in Section 6.2.2.

We can repeat almost exactly the same formulation of the Second Law for this process (compare with Section 6.2.2).

> When two games of equal sizes, each at equilibrium having different values of *TA*, are brought to a new equilibrium state, under the constraint that total cloth area is conserved, cloth will always flow from the system having a higher value of *TA* to the system having a lower value of *TA*. The value of the *SMI* of the combined system at the new equilibrium state will be larger than the sum of values of the *SMI* of the two systems in the initial state.

We note again that this formulation of the Second Law applies to two systems of equal sizes (i.e. same *NM* and *NC*), but different in their *TA* value.

The answer to the question of why the system will proceed to the new equilibrium state is exactly the same as that provided in Section 6.2.2; namely, that the probability of the final equilibrium state will be larger than that of the initial state. The probability of the combined system is proportional to the product of the multiplicities of the two systems. This product of the multiplicities can be read as the *sum* of the logarithm of the multiplicities from Fig. 4.26 or Fig. 4.29.

6.3. Summary of Everything We Have Learned: What, Why, and How

This is a summary of all the summaries of Sections 6.1–6.2. In all the experiments we carried out, whether on a single system or a pair of systems (and by generalization on any number of systems), we have observed some changes in the state of the system. This could be a flow of marbles from cell to cell within the same system, flow of marbles between two systems, flow of string between two systems, or flow of cloth between two systems. At the most elementary level, the change that took place was in the *state* of the system. This change was such that after a long shaking of the system, it reached a new equilibrium state characterized by a maximal value of *SMI* for that system under whatever constraints we have imposed. We have also seen that the change is characterized by maximal value of *W*.

Thus, *what* had changed was the state of the system (we always follow the dim-state). *What* characterized the change in the system was the increase in the value of

SMI (until equilibrium is reached, when it then stayed constant). Why did it change in that direction? Because the probability of the final state is always higher than the probability of the initial state.

In this chapter, we have made several formulations of the Second Law for the different experiments. This is a reflection of the multitude of manifestations of the Second Law. What is common to all these formulations, as well as the ones in the real world, is that whenever a spontaneous process occurs, it always involves an *increase* in the *SMI* of the entire system.[10] Why does the spontaneous process occur? Because the *SMI* is also related to $\log_2 W$ and W is related to the probability of the state of the system. Thus, in all cases, a spontaneous process is associated with an increase in *SMI*, or what can be called the *M*-entropy of the system. The processes occur spontaneously, because the new equilibrium system has a higher probability relative to the initial equilibrium system.

There remains only one further question to answer: *How* — how did we get from the initial to the final state?

The answer to this question depends on how you performed the experiment.

If you performed a real shaking of the system of marbles, then the physical *shaking* is the answer. If you performed a simulation of the experiments on the computer, then it is the program you have used to achieve the final state of equilibrium. This is what we have done in Chapters 3–5.

If, on the other hand, you are dealing with a system of real atoms and molecules evolving from one state to another, then it is the kinetic energy of the atoms and molecules that enables the system to move from one state to the other, but this answer belongs to the next chapter. Here, we were concerned only with games of marbles in cells.

Snack: The Mystery of the Missing Bike and the Power of the Ten Commandments

Mayor Don T. Still would walk to work each and every single day. His office, in the City Hall, was a stone's throw away from the parish church. Each morning, he would meet Father Lye N. Cheet along the way, who was always merrily pedaling his bike, whistling a happy tune and who seemed not to have the slightest worries.

One cold and windy winter morning, Mayor Still saw Father Cheet, for the first time without his bike. Gone was the priest's usual cheerful demeanor and what the Mayor saw instead was a man who looked forlorn and defeated. With slumped shoulders, the priest dragged his feet and walked towards the mayor. "Good morning, Father," the mayor greeted the priest. "What happened to you? Why do you look so tired? Where is your bike?"

"This morning, I got up early as usual, and as I was getting ready to leave, I discovered that my bike, which I had never locked before, was missing. That bike means so much to me as it was my mother's present on my ordination. So today I had no other choice but to walk on this cold and dreary morning," the priest sighed.

"That is quite serious. I pride myself as mayor of this city for a crime rate of almost zero percent since I took over. I promise you that I am going to use everything in my power to find out who stole your bike. But tell me, do you have any suspects?"

"Not at all," answered Father Cheet.

"I have a suggestion to make and it might just work out. You can make an announcement to your congregation to come and hear the morning Mass this coming Sunday. Tell them that you have something special to announce. I have attended several of your Masses and I must admit I was amazed at how you manage to translate the readings into very relevant, day-to-day stories which the parishioners could relate to. I suggest that you talk about the Ten Commandments. You read all the commandments slowly, and when you get to the commandment about stealing, you look everyone in the eye. For emphasis, I suggest you repeat it several times: "Thou shalt not steal, Thou shalt not steal". If one of them suddenly turns jittery and uneasy, then you would know who the thief is," the mayor said confidently.

"That's a brilliant suggestion. I will do exactly as you told me."

The following week, on Monday morning, while the mayor was walking on his way to the office he saw the priest riding his bike as usual and it seemed as if all his worries had dissipated. He even had that radiant glow on his face.

"Good morning Father, I see that my advice worked," the mayor said.

"Oh yes it did! And I cannot thank you enough for that brilliant suggestion," the priest said hurriedly. "But tell me Father, how did it go? Did you actually catch the thief? What do you plan to do now? Shall we put him behind bars?"

"Oh no, no, no! It's alright, everything is alright now." He was obviously embarrassed and trying to evade the mayor's questioning, when the mayor stopped him in his tracks and asked: "But how, tell me, how did you catch the thief?"

"There is no criminal," the priest said, trying to avoid the mayor's eyes and mounting his bike. "Huh? How can that be possible?" the mayor asked.

"I followed your instructions, exactly as you told me to. I read each and every commandment slowly and loudly: "Thou shalt not kill; Thou shalt not commit adultery ..."

"At that very moment, when I uttered those words, I remembered where I left the bicycle," the priest said, and started to pedal away without giving the mayor the chance to speak.

Happy days...

A sad day...

Happy days are here again...

Snack: The Ever Reliable Cuckoo Clock

Bill O' Flirty was happily humming a tune to himself on the way home, when reality set in. "Oh no! It's almost 1 a.m. I hope Monique did not wait up for me."

Fumbling for the house keys in his trouser pocket, and finally opening the front door, he said to himself: "Be quiet, Bill. You do not want to wake up Monique."

As soon as he entered the house, he took off his shoes and, with calculated steps, descended the stairs. As he opened their bedroom door, he tiptoed stealthily and was relieved to see Monique's silhouette lying sideways on their bed.

His relief was short-lived, however. He accidentally stepped on Monique's wooden backscratcher lying on the floor. This noise roused Monique.

"Bill, is that you?" she asked, half-asleep. "Why did you come home so late?"

"Late? Darling, it's only 10 o'clock. Go back to sleep, darling." After being reassured that it was only 10 o'clock, Monique sleepily murmured "Goodnight darling."

Just as Bill thought he had gotten away with it, the ever reliable cuckoo clock went "ding dong" once and then silence.

Suddenly, like a raging bull, Monique got up and said: "You said it's only 10 o'clock, Bill. But the cuckoo clock never lies. The alarm went once only, which means it is 1 a.m. and not 10 o' clock".

"Darling, of course the clock went off only once — don't you know that the clock never sounds off the zeros?" Bill asked defensively.

Bill was correct, and what he said was true. The clock's alarm never sounds on zeros. But was he truthful?

CHAPTER 7

Entropy and the Second Law in the Real World

After long and tedious experiments with strange games, and after careful analysis and seemingly profound conclusions reached on such nebulous games that no one has ever played, you must be weary by now and worry about the relevance of all these piles of data and the sophisticated conclusions to the real world. Why should I care for a game which, whenever it evolves, becomes more difficult to play?

My answer is that you are right! Indeed, these games are complicated, uninteresting, and certainly no one has ever played them, let alone expended an effort to analyze their "behavior".

However, in my defense I would say that although these games are in themselves irrelevant to anything real, they can be translated into very real processes and the conclusions are nothing but — as you can expect — the Second Law of Thermodynamics.

The reason that I have "invented" all of these unrealistic games is that I believe that using the language of games, even imaginary games, will be more attractive to you. If I were to start with systems of huge (unimaginably huge) numbers of atoms and molecules, with kinetic and potential energies, if I were to discuss the entropy of a system constrained to have a fixed total energy, and so forth, you would probably set this book aside.

In choosing the game language I hope I made you more comfortable, simply because I have never mentioned any concept that you are unfamiliar with. The most sophisticated concept I used is SMI, but I used this as a short-hand notation for the ANOQONTAITSS, which was an acronym for "average number of questions …".

I fully agree that the experiments we have done were tedious and seem to be aimless. But as I assured you in the Preface, working through these experiments will be rewarding. It is now time to fulfill my promise and to reward you for being with me throughout this long journey. This chapter is about the real world, about the real entropy, and the real Second Law of Thermodynamics.

Reading through this chapter you will understand why I made such an effort to invent all these strange games. You will also be rewarded by understanding what entropy is and what the Second Law is all about.

To reach that goal, all you have to do is to translate from the language of marbles and cells to the language of real atoms and molecules. If you have understood the conclusions of Chapter 6 then you must be able to understand the conclusions in this chapter, and you will also understand my motivation for inventing these games.

The first section of this chapter is the dictionary. Once we adopt the new language we shall simply translate each one of the conclusions from Chapter 6 into the language of real particles.

7.1. The Dictionary

Here is the complete dictionary that you will need for translating from the terms used in the world of games to the real world.

Marbles → atoms and molecules.

Cells → either cells, or boxes, or any container in which the atoms and the molecules are contained.

NM → number of independent particles. For simplicity we can think of the number of argon atoms in the gas phase. The numbers of particles in the real world are very much larger than the numbers of marbles we have discussed in the games.

NC → Number of cells or boxes. These could be very large or very small depending on the system we will discuss.

Specific state → A list of the locations and/or velocities of *each* specific particle.

Dim-state → A list of how many particles are in certain locations and having certain velocities.

Equilibrium state → A state of the system when we do not notice any change in any measurable quantity at each point in the system (e.g. density of particles at each point, or the average number of particles having velocities between, say, 100 m/s and 110 m/s).

TL → The total energy of the system.

TA → The total kinetic energy of all the particles — this is a particular case of the total energy, when all the energy is in the form of kinetic energy.

Shaking → Here, we do not need to shake the system. The "shaking" comes from *within* the system. The random motion of the particles causes the system to go from the initial to the final equilibrium state.

W → Exactly the same as in the game. This is the number of specific states of a system belonging to a given dim-state.

SMI → You might have guessed that I will translate *SMI* into entropy. You are right. Nevertheless, I want to retain the same term *SMI* even for the real world. As I shall explain later, entropy is a particular case of *SMI*. The inverse of the last sentence is not true. *SMI* is a far more general concept than entropy and, in most cases, very different from entropy.

Therefore, I shall keep using the term *SMI*, which I believe is a more adequate term to use, even for a real system. I shall indicate in the following sections when the *SMI* becomes identical with the more commonly (but in my opinion inadequately) used term "entropy."

With this list of terms we can start our task of translation. I know you might suspect that I have missed one important term — the temperature. Indeed, it is an important term, and I shall discuss that later in this chapter.

7.2. Uniform Distribution: Shannon's First Theorem and the Second Law

In this section, we shall discuss the processes of expansion of a gas. We shall always assume that the molecules do not interact or, if they do, that the interactions are insignificant. Such a system is called an ideal gas. This is a very convenient model to use to study the behavior of real gases at very low densities or pressures. In this section, we shall discuss in detail *what* happens, *why* it happens, and *how* it happens.

7.2.1. *What Happens?*

We start with the simplest experiment we have carried out in Section 3.3. Reminder: We started with *NM* marbles in one cell, and no marbles in the second cell. We

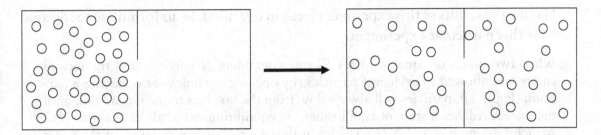

Fig. 7.1 Expansion of a gas.

"shook" the system and found that the system had changed gradually from the initial state to the final equilibrium state. Along the way, the *SMI* increased steadily towards a maximal value; so did *W* and *PR*. The equilibrium density profile was such that about half the marbles were in each of the two cells.

The translation: We start with *N* particles in one box and an adjacent empty box (Fig. 7.1). The system does not need shaking. It is already vigorously shaken from *within* the system. The particles are moving and jittering at high speed, occasionally hitting the walls of the container. We open a small window between the two boxes and allow the system to evolve using its own power. What will happen?

First, let me note that the "smallness" of the window is not essential. We could have opened a big window, or even removed the partition between the two boxes. I chose a small window for a later discussion of the step-by-step evolution of the system. At this moment, the size of the window is inconsequential.

As you have seen in the experiment in Section 3.3, the system will evolve in such a way that particles will flow from the first box to the second box. Along the path from the initial to the final equilibrium state, the *SMI* will gradually increase until it reaches a maximal value. The same upward thrust was observed for the values of *W* and *PR*. We also followed the number of particles in the first cell, starting from the initial value of *N* in the initial state to the value of about *N*/2 in the final state.

You have already observed all these *facts* in the experiment of Section 3.3. You can therefore claim that the same thing must be observed in the real system described in Fig. 7.1. The only thing that you have to accept based "on faith" is that you do not need to *shake* the system since it is already incessantly shaken from *within*.

Having the results of this experiment fresh in our mind, let us formulate the Second Law for this particular experiment:

When two boxes of equal volumes V, one containing N particles and the second empty, are allowed to exchange particles (by opening a window or removing a partition (Fig. 7.1)), particles will always flow from the first box to the second box until the system reaches a state of equilibrium. At equilibrium we shall have about $N/2$ particles in one box and $N/2$ particles in the other. If N is very large, of the order 10^{23}, we shall not see any deviations from this equilibrium state.

We now have a statement of the Second Law of thermodynamics applied to a real system of real particles in real boxes.

Of course, we can say much more about this process than merely reckoning the changes in the number of particles in each of the boxes.

First, I have restricted the Law for the case of two boxes of equal volumes. This restriction will be removed soon.

Second, one can prove that the formulation of the Second Law as stated above is equivalent to any other statement of the Second Law. I do not have to bother you with that proof now. If you are interested you can look at any textbook on thermodynamics.[1]

Third, the statement was made regarding the dim-states and not the specific states. In the games of marbles we could, in principle, label the marbles by numbers (or by colors, or whatever) and describe a *specific* state of the system; that is, in which cell each specific marble is. In the game of marbles we voluntarily moved from the specific description to the *dim* description, simply by removing the labels. In the real world, we cannot *label* the particles. Therefore, we cannot see the *specific* states. All we can see, record, and follow is the dim-states; that is, how many particles are in each box. The particles are, in principle, indistinguishable.

Fourth, when we followed the *SMI* of the system of marbles, we found that the *SMI* of the system changes from $SMI = 0$ (no questions asked) to $SMI = \log_2 2 = 1$ (one question needed to locate the cell in which a specific marble is). This change of *SMI* was for one marble. If we have to find the location of NM marbles, then the corresponding *SMI* will be $NM \times \log_2 2 = NM$.

Exactly the same is true for the change in the *SMI* that we would have followed during the expansion of the particles from one box to the other. Initially, we know that each of the particles is in the first box. At the new equilibrium state each particle may be in either of the two boxes. Since the boxes are identical, the probability of finding any specific particle in either the first or the second box is 1/2. Therefore, the *SMI* per particle is $\log_2 2 = 1$ and the change in the *SMI* in the process for N particles is $N \log_2 2 = N$.

At this point, you might have asked yourself why I elude mentioning the change of the entropy of the system. To assuage your apprehension, I shall discuss it explicitly, although I have just discussed that issue.

Before discussing entropy you should know that entropy is defined only for equilibrium states, and that in thermodynamics only *changes* in entropy are discussed and not absolute entropies.

Having made these two comments, let me say the following. The *SMI* is a quantity that may be defined for each dim-state, whether of a small or large system, whether at equilibrium or not. In the particular process of expansion that we have discussed in this section, the change in *SMI* could be followed along the entire path from the initial to the final state. However, for the net change between the initial and the final state, the change in the *SMI* is the same as the change in the entropy of the system.

Now you can see why I have retained the term *SMI* and did not translate it into entropy in the dictionary. The *SMI* is a much more general concept than entropy. It applies to the 20Q games, to the marble games, to states at equilibrium or non-equilibrium. In general, the *SMI* is not entropy. On the other hand, the entropy is a particular case of an *SMI* applied to a system of real particles in a box and at equilibrium. The change in the entropy from one equilibrium state to another is the same as the change in *SMI* between these two states.

It should be noted that once we open the window or lift the partition between the two boxes, the system at that point is not at equilibrium. All we can claim about the difference in the entropy is between the initial and the final equilibrium states. If we want to examine the evolution of the *entropy*, not the *SMI* of the system, from the initial to the final state, we must open and close a small window, wait until the system equilibrates, determine the change in entropy, and repeat this process. This type of process is called a quasi-static process. It consists of an almost continuous series of equilibrium states. In such a process it is meaningful to talk about entropy change along the entire process.

There is one incidental difference between changes of *SMI* and changes of entropy. These are the units used in the two cases. The *SMI* is measured in units of bits, whereas entropy is measured in units of energy divided by the temperature.

The fact that the entropy got these units is a historical accident.[1] We shall discuss this matter further in Section 7.4. Here, we shall say only that neither energy, nor temperature, is necessary to calculate entropy changes. This brings me to the second concept that I did not mention in connection with the formulation of the Second Law: the temperature. I did not mention temperature because temperature does not feature in the change of either *SMI*, or the entropy in the process of expansion of independent particles. All that matters is that each particle was initially in one box and finally in either one of the two boxes. The change of the *SMI*, as well as the change in entropy, is the *difference* in the *SMI* between the initial and the final states, which in our example is simply $\log_2 2 = 1$ per particle.[2] There is no temperature and no energy involved in this process. This is only one example which demonstrates unequivocally that involvement of energy and temperature in the Clausius definition of entropy is not essential.

I should note however that the kinetic energy of the particles is involved in the *shaking* of the system, and enabling it to move from the initial to the final states, but not in the *determination* of the *value* of the entropy change. This is a very important point. In the expansion process, the kinetic energy of the particles is only involved in the "*shaking*" — which enables the *changes* in the *locational distribution* of the particles. In Section 7.4, we shall discuss another important process; the heat transfer from a hot to a cold body. In the latter process, the kinetic energies of the particles play a double role, in the "shaking" of the system; that is, in enabling *changes* in the *velocity distribution*, as well as in the *velocity distribution* itself. I shall expound on this in the next sections.

7.2.2. Why it Happens?

It is often said that processes occur spontaneously *because* the entropy increases in that process. It is true that in any spontaneous process (occurring in isolated systems) the entropy increases. We shall see many such processes in this chapter. However, the change in entropy cannot be used to *explain* why the process has occurred. Note that in our particular formulation of the Second Law on page 194, neither entropy nor SMI was mentioned. We have described *what* has happened, but not *why* it has happend. Although the Second Law says that the entropy increases in such processes, it does not offer an explanation as to why the entropy increases, or why the processes occurred spontaneously.

The answer to the question of *why* is probabilistic. If we accept the relative frequency interpretation of probability, then states which are more probable will be visited more frequently. Since the probability of each dim-state is proportional to the multiplicity (W) of that state, we can say that the larger the multiplicity, the larger the probability of that dim-state.

In the experiments that we have carried out in Chapter 3 (as well as in Chapters 4 and 5), we observed that both the *SMI* and the *W* reached a *maximal* value at equilibrium. Maximal value of *W* means maximal probability of that dim-state. This explains why the system will proceed from the relatively lower probability state to the higher probability state. In other words, the relative frequency of visiting the higher probability states will be larger. This still does not explain why a system with large number of particles will *always* stay in the equilibrium state.

To explain this we need the law of large numbers.[1] Qualitatively, we must recognize that the equilibrium state of a system is not a single dim-state but a small group of dim-states near the dim-state, having the maximal multiplicity. If we calculate the probability of occurrence of this group of states, we find that it is nearly one. Probability of "nearly one" means that the system will almost certainly be in one of the states of this group. For a very large number of particles, probability "nearly one" means that the system will, in practice, *always* stay in that group of states.

Because of the utmost importance of the expansion process for understanding the Second Law (the second important process will be discussed in Sections 7.4 and 7.5), it is instructive to follow the expansion process step by step.

Suppose that we start with a system of N particles in one box of volume V as in Fig. 7.2, but the second box is initially empty. Now, we open a tiny window to allow the flow of particles one at a time. Because of the random motion of all the N particles in the volume V, and because the particles are identical, each *specific* particle has the same chance of crossing from the left (L) compartment into the right (R) compartment in some small interval of time (we can always choose the interval

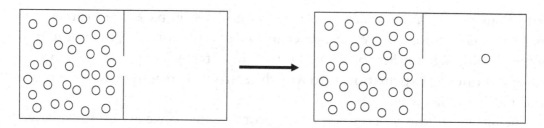

Fig. 7.2   Step-by-step detailed consideration of the expansion of a gas.

of time t small enough that at most, one particle can cross the window in that period of time).

Let us denote by p the probability of a *specific* particle crossing from L to R in the time period t. If there are N particles in L, then the probability of *any* particle crossing from L to R in this time period is simply $N \times p$. Once the first particle has crossed we can repeat the calculation of the probabilities of crossing from L to R and from R to L.

Since there are now $N - 1$ particles in L and only one particle in R, the probability of *any* particle in the system crossing from L to R is $(N - 1)p$, but only p from R to L. If N is very large, the chances of crossing the window will be overwhelmingly in the $L \rightarrow R$ direction. We can continue with the same argument for any dim-state — say, when we have $N - 20$ in L and 20 in R. The probability of crossing from L to R is $(N - 20)p$ but only $20p$ from R to L.

Clearly, as long as the number of particles in L is much larger than the number in R, the probability that a particle will cross from L to R will be larger than the probability of crossing from R to L. Therefore, over a longer period of time we shall observe a *net flow* of particles from left to right. This will continue until about half of the particles, $N/2$, have crossed from L to R. At this stage the probability of *any* particle to cross from L to R will be almost the same as the probability of crossing from R to L (i.e. $(N/2)p$). At this stage, particles will continue to cross the window in both directions. However, *on average*, the number of particles that flow from L to R will be roughly the same as the number of particles that flow from R to L. This means that there will be no *net flow* of particles in either direction. Thus, on average, the number of particles will remain at around $N/2$. Of course, there will be small fluctuations about the number $N/2$, but these fluctuations are either too small to be observed, or large but too improbable and therefore unobservable. This is the state of equilibrium where we observe no changes in the system. It is important that you repeat the argument we have just presented. Although the argument is only qualitative, it is essentially correct and it shows why the process of expansion will always proceed in one direction only. Once the equilibrium state is achieved, the system will remain there for a very long time. For a very large number of particles, "very long time" is practically "forever." In Section 7.5, we shall discuss another one-way process. The argument there is a little more difficult, but we already have all the necessary tools to be able to grapple with that process, too.

7.2.3. *How it Happens?*

Having answered the questions of *what* entropy is and *why* it changes in one way only, we should devote some time to the question of *how*.

First, and foremost in importance, the answer to the question of *how* is trivial. In all the experiments mentioned in this book, the answer to the question of *how* the system proceeded from one state to another is simply the "shaking." The shaking was mechanical in the case of marbles in cells. In a real thermodynamic system, the shaking is from the "inside" — that is, by the molecular motions of the particles. In the simulation, the shaking was achieved by the program of selecting a particle at random and placing it in a randomly selected cell. There is no mystery in either of these.

Second, thermodynamics does not deal with the *way* a system moves from one state to the other. It was long ago suggested that the term thermo*dynamics* should be replaced with thermo *statics*. The entropy of a system is defined for a system at *equilibrium*. Thermodynamics is concerned only with the *difference* in the entropy of a system between two equilibrium states, and not with the *way* or the path along which this change of state has occurred. This is exactly what is meant when referring to the entropy as a *state function*. It is a quantity, the value of which is determined by the *state* of the system and not by *how* the system got to that state.

At this point, I have given you an answer to the *how* question, and assured you that this question, as well as its answer, is irrelevant to thermodynamics. The answer is both trivial and unnecessary. It does not shed any light on the mystery encompassing the entropy nor on the meaning of the Second Law of thermodynamics. It should be noted that the literature (as well as the Internet) is replete with statements that intertwines the "internal shaking" of the system with the concept of entropy. Some authors even suggest that focusing exclusively on probability considertaions in entropy change is flawed, since it ignores the molecular motional energy. Such statements reveal a profound misunderstanding of the every concept of probability.[3] For more details see Ben-Naim (2008) Chapter 6.

7.3. Some More Processes in the "Uniform" Class

In the previous section, we discussed one of the most important processes in which the Second Law was involved, and for which we could clearly identify the meaning of entropy. The second most important process will be discussed in Section 7.5.

In this section, we shall discuss very briefly some other processes related to the expansion process. All of these appear in textbooks on thermodynamics, sometimes in insufficient detail, sometimes in distorted or confusing ways.

7.3.1. *An Alternative View of the Expansion Process*

In Chapter 3, we discussed two expansion processes. The analog of the first (Fig. 3.21) was discussed in Section 7.2. Here, we will discuss the analog of the second (Fig. 3.24). The analysis of both processes is important for reasons which will be made clear below.

Consider two compartments each of volume V and each divided into M cells. Initially, we have N particles in the left-hand side compartment. We prepare the system by placing N_1 particles in cell 1, N_2 particles in cell 2, and so on until we reach N_M particles in cell M. The total number of particles in the compartment is $N = N_1 + N_2 + \cdots + N_M$.

The particles are assumed to be small enough to be non-interacting with each other. Therefore, there is no limit on the number of particles in each cell. One simple choice of an initial state is $N_1 = N$, and all other cells being empty (Fig. 7.3a). This will be the analog of the system described in Section 3.3.

We first bring the system to equilibrium by "shaking" it, as we did in Chapter 3. The only difference is that here the "shaking" is done from the "inside." We allow

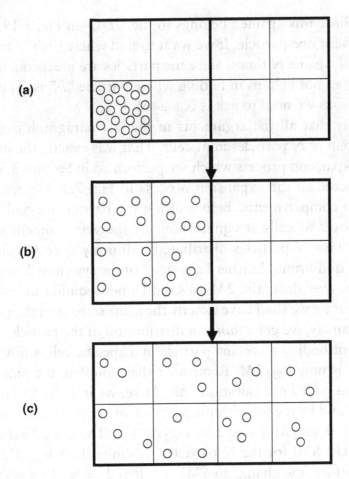

Fig. 7.3 Expansion of a gas. The same as in Fig. 7.1 but now the gas is initially distributed in one of *M* cells.

the particles to cross freely from cell to cell, but not to cross over to the right-hand side compartment. We know what the result will be: Particles will move randomly from cell to cell until at equilibrium — a *uniform* distribution of particles in all the *M* cells will be reached. The average number of particles per cell is now N/M. The final outcome has a maximum value of *SMI* and a maximum value of *W*. More specifically, in terms of the 20Q game, we shall need to ask $\log_2 M$ question per particle. This follows from the fact that, at equilibrium, the distribution of particles in the cells is uniform, therefore the probability of finding a specific particle in a specific cell is simply $1/M$. Since this "game" belongs to the *UDG* (in Fig. 2.19), the *SMI* value is $\log_2 M$. That is for one particle. If we want to find where each of the N particles is, we have to play the game N times. Since the particles are independent, finding where particle "k" is does not help us in finding where particle "h" is. Therefore, the total number of questions we need to ask is $N \log_2 M$.

Note carefully that all the arguments in the last paragraph are correct for the equilibrated system of N particles in M cells. That was exactly the initial state of the system for the expansion process which we performed in Section 3.3.

We now proceed to the expansion process in Fig. 7.3. We remove the barrier between the two compartments. Before lifting the barrier, we had N particles distributed evenly over M cells at equilibrium. At the very moment we removed the barrier, we still have N particles distributed uniformly over M cells. But now the system is not at equilibrium. Further "shaking" of the system will result in spreading of the N particles over the entire $2M$ cells until a new equilibrium is reached. At the new equilibrium state we shall have exactly the same state as in the process discussed in Section 7.2; namely, we get a *uniform* distribution of the particles in the $2M$ cells. The probability of finding a specific particle in a specific cell is now $1/2M$, and the *SMI* per particle is now $\log_2 2M$. Remember that doubling the number of boxes in the 20Q game has added one question only. Here, we initially had $\log_2 M$, and now we have $\log_2 2M$ and by the logarithmic property of the product of the two numbers, we have $\log_2 2M = \log_2 M + \log_2 2 = \log_2 M + 1$. Thus, the *SMI* per particle has increased by 1. The *SMI* for the N particles was initially $N \log_2 M$, and finally it is $N \log_2 2M$. Therefore, the change in *SMI* is $N \log_2 2 = N$. This is *exactly* the same result we have obtained in Section 7.2. We can conclude that the change in the *SMI* in the expansion process is N, *independently* of the number of cells. We could divide the volume V into any arbitrary number of cells. The *SMI* of a system of N particles in M cells *depends* on M. This is $N \log_2 M$. But the *change* in the *SMI* in the process of expansion from M cells to $2M$ cells is *independent* of M. The *reason* for the spontaneous expansion process in both cases is the same; going from a low probability to a high probability state.

We could have reformulated the Second Law for this process with exactly the same words we used in Section 7.2, without mentioning the number M. We can also identify the change in *SMI* from one equilibrium state to another with the change in the *entropy* of the system in this process. In thermodynamics we use other bases for the logarithm and other units for the *SMI*, but the choice of bases and units is incidental. The entropy change in the process of expansion from volume V to $2V$ is written as $N \log(2V/V)$, which for base 2 is simply $N \log_2 2 = N$. In both cases, the eventual density distribution is uniform, a result which follows from Shannon's first theorem.

In this section, we have learned one important lesson: The *entropy* or the *SMI* of a system *depends* on how we choose to divide the volume into cells. However, the *change* in the entropy is *independent* of such a division into cells (provided of course that we use the same cell sizes in the two compartments).

7.3.2. *Pure Mixing*

The following is a well known process of the mixing of two gases — say, argon and neon. Initially, each gas is in a separate box of volume V. The final state is a mixture of the two gases in the same volume V (Fig. 7.4). We have seen in Chapters 3 and 6 that in this process, the *SMI* does not change. We can also conclude that in the process depicted in Fig. 7.4, there will be no change in the *SMI*, hence no change in the entropy. The rationale for this result is simple. In the initial state, we have some value of *SMI* for each type of gas. Let us say we divided the volume V into M equal cells. Then the *SMI* per argon atom is simply $\log_2 M$. And for the neon atom it is also $\log_2 M$. In the final state, we also have the same *SMI* for each argon atom and for each neon atom. Therefore, there should be no change in the *SMI*. That is what we found in Sections 3.3 and 6.2 and that is what is known experimentally for a real mixing of two gases. Under these conditions there is no change in entropy.

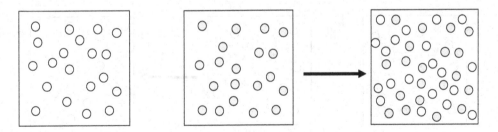

Fig. 7.4   A pure mixing process.

One caveat before moving on to the next process: Entropy is most often associated with a degree of disorder. Mixing is viewed as a process in which the disorder increases. Therefore, it is very tempting to conclude that mixing is associated with an increase in disorder. That conclusion is unfortunately wrong, as the example discussed in this section demonstrates.

7.3.3. *Pure Assimilation*

The process of pure assimilation is almost the same as pure mixing, except that the process involves two identical gases instead of two different gases. The process is depicted in Fig. 7.5. Initially, we have N particles in a box of volume V and another N particles of the *same kind* in another box of the same volume V. We "mix" the two gases to form a system of $2N$ particles in one volume V.

Clearly, the entropy change in this process is negative; more specifically, the change in the *SMI* is -1 per particle.

To see this, suppose we divide each of the volumes into M equal cells. The *SMI* per particle in the initial state is $\log_2 2M$ (there are altogether $2M$ cells in which the particle could be hidden). In the final state, the *SMI* per particle is only $\log_2 M$. Therefore, the change in *SMI* per particle in the process is $\log_2 M - \log_2 2M = -\log_2 2 = -1$, and for $2N$ particles is $-2N$.

Thus, we conclude that in the pure assimilation process the entropy decreases. Don't rush to conclude that you have found a process where the entropy decreases. There are many processes we can think of in which entropy decreases. The Second Law states that in a spontaneous process in isolated systems, the entropy always increases. Here, the combined system is isolated, but the process is not a spontaneous

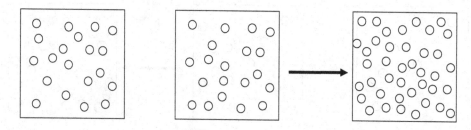

Fig. 7.5   A pure assimilation process.

one. There exists a reverse of this process referred to as de-assimilation, which is a spontaneous process and for which the entropy increases. See Section 7.3.6.

A final caveat for this section, too. We have calculated the change in *SMI* in the process of assimilation as if the volume has decreased from $2V$ to V (or from $2M$ cells to M cells). The result is correct but there is a subtle conceptual point here. In the statistical thermodynamics treatment of this process, the volume per particle does not change in the process, but only the number of indistinguishable particles has changed. For more details see Section 7.3.5 and Ben-Naim (2008).

7.3.4. *Mixing and Expansion*

This experiment was done in Section 3.3. We saw that it is equivalent to two processes of expansion. It is also true that this process of mixing of the two real gases is completely equivalent to the two processes of expansion. We start with N atoms of argon in one compartment of volume V, and another N atoms of neon in a second compartment of volume V. We then remove the barrier between the two compartments. What will happen? The argon atoms from one compartment will expand to occupy the two compartments, and so will the neon atoms. The final equilibrium state will be a uniform distribution of both the argon particles and the neon particles in the entire extended volume of $2V$ (Fig. 7.6a).

Since *there is nothing new* in this experiment, you can re-read Section 7.2 twice, one for a system of N argon particles and one for N neon particles, and make your own conclusions as to the change in entropy and the change in the probability of the combined system when it evolves towards the final equilibrium state.

I said "there is nothing new," and I meant to say that *there is nothing new* in this experiment. It is completely equivalent to two experiments of expansion! Nevertheless, open any textbook on thermodynamics or statistical mechanics and look in the index under "entropy of mixing" or "mixing, entropy." You will find a few pages dedicated to the discussion of the "entropy of mixing." This discussion appears in addition to, and independently of, the expansion process. This is quite puzzling since we have just concluded that the process described in Fig. 7.6a is simply another example of an expansion process. You might also ask: Why did *I* devote a separate section to discussing "mixing and expansion" when I had just said that this process is the same as the process of expansion?

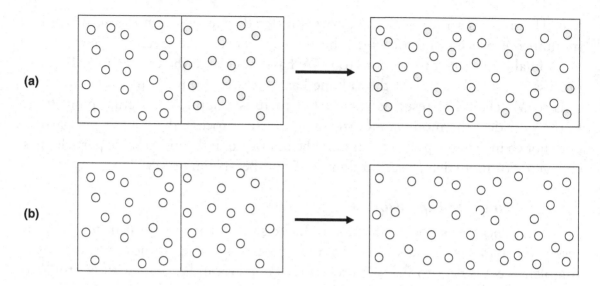

Fig. 7.6   (a) A mixing and expansion process. (b) An assimilation and expansion process.

As always, you are perfectly right. I should not have written this section at all!

Nevertheless, I found it necessary to *write* this section to inform you that such a section *need not be written*, either here or in any textbook on thermodynamics.

The reason authors do discuss the "entropy of mixing" and even conclude that "mixing is an inherently irreversible process," and that "mixing increases disorder," and "increase in disorder is equivalent to increase in entropy," and so on, is a result of the misunderstanding of the process of mixing and expansion. The source of this misunderstanding goes back to Gibbs,[4] who was the first to analyze this process over 100 years ago. Since then, many people associate mixing with increase in disorder (which, on qualitative grounds seem reasonable), and that disorder is associated with entropy. Therefore, one concludes that the *mixing* must be the cause of the change in the entropy in the process described in Fig. 7.6a, and therefore the increment of the entropy in this process is called the "entropy of mixing."

Unfortunately, that is not true. We have already seen a *pure* mixing process with no entropy change at all. Furthermore, one can devise an experiment where *demixing* is spontaneous, followed by an increase in entropy. Clearly, one cannot conclude that both mixing and demixing causes an increase in entropy! For more details on this, the reader is referred to Ben-Naim (2008).

I should also add that the reason for misunderstanding the process of "mixing and expansion" as described in Fig. 7.6a stems from misunderstanding another process referred to as "assimilation and expansion", which I will discuss in the next section.

7.3.5. *Assimilation and Expansion*

Look at Fig. 7.6. There are two processes — the "mixing and expansion" which is equivalent to the process described in Fig. 7.1, and another one which is, well, what to call it? Perhaps a "non-process." After all, nothing conspicuous happens when we remove the barrier between the two compartments in Fig. 7.6b.

Looking at the two processes in Fig. 7.6, you already know that in the first one, the entropy change is $2N \log_2 2 = 2N$, and in the second one there is no change in entropy. Therefore, it is obvious that this process may be referred to as a "non-process," right? Not really!

Indeed, when we remove the barrier between the two compartments nothing conspicuous happens. There is no change in entropy. However, it is not true to say that *nothing* is happening. As we lift the barrier, particles from one compartment will wander into the entire extended volume $2V$, as do the particles of the second compartment. Even Gibbs who first analyzed these two processes said that the two gases of the *same* kind will be *mixed*; that is, inter-diffuse into each other.[4]

So something does happen, although it is not visible. That is why I called this process "assimilation and expansion" and not a "non-process." Visually, the process in Fig. 7.6b seems to be simpler and we deem it to be a "non-process." However, conceptually the simpler of the two processes in Fig. 7.6 is the *mixing* process, Fig. 7.6a, and the "non-process," Fig. 7.6b, is not simple at all.

Here is a quotation from Gibbs himself.[4]

> If we should bring into contact two masses of the same kind of gas, they would also mix but there would be no change of energy or entropy. We do not mean that the gases which have been mixed can be separated without change to external bodies. On the contrary, the separation of the gases is entirely impossible.

In simple words, what Gibbs is saying in that paragraph is that in the process in Fig. 7.6a, which is a spontaneous process of *mixing*, one can bring all the argon atoms and all the neon atoms to their original compartment. This will require some work to invest. However, in the process in Fig. 7.6b, the separation of the gases (of the same kind) *cannot* be achieved. No matter how much energy we use, the reversal of the process in Fig. 7.6b is "entirely impossible."

To summarize, looking at the two processes in Fig. 7.6, we *see* mixing in Fig. 7.6a and we see that nothing happens in Fig. 7.6b. We know that the first process involves positive change in entropy, and no change in entropy in the second process. Therefore, it is very "natural" to conclude, as Gibbs and numerous others did, that the *mixing* depicted in process Fig. 7.6a is the *reason* for the increment of entropy and obviously the "non-process" in Fig. 7.6b involves no change in entropy.

The facts are correct, but the interpretation of the facts is not.

In the mixing process of Fig. 7.6a, the cause of the change in entropy is nothing but the *expansion*. In the apparent "non-process" in Fig. 7.6b, there are actually *two* processes that happen simultaneously. Each molecule that was confined initially to a volume V can now access a larger volume $2V$. In addition, each particle was initially a member of N indistinguishable particles. After the removal of the barrier, it is a member of $2N$ indistinguishable particles. Each of these two processes affects the entropy in such a way that the first causes an increase in the entropy while the second causes a decrease in the entropy. It turns out that the two effects cancel each other and the net effect is a zero change in entropy. This is the reason I have referred to this process in the title of this section as "assimilation and expansion." We have seen that the assimilation causes a decrease in entropy and the expansion causes an increase in entropy. To show that these two changes in entropy cancel each other needs some mathematics. A short comment is provided in Note 5, and a more thorough discussion in Ben-Naim (2008).[6]

When Gibbs reached the conclusion that reversal of process in Fig. 7.6b is "entirely impossible," he probably had *mentally followed* the process in Fig. 7.6b by *mentally labeling* each particle and following its trajectory from the initial to the final state. Then he realized that there is no way that one can bring back each particle to its original compartment. Why? Because the particles are indistinguishable. Therefore, there exists no contraption that can *separate* all those particles that originated from the left compartment from those that originate from the right compartment.

Gibbs was right that such a contraption does not exist. But Gibbs failed to realize that exactly for the *same reason* he reached the conclusion "entirely impossible," the reversal of process Fig. 7.6b is *trivially possible*. Simply place the partition in its original place and you are back to the initial state. There is no way one can distinguish this state from the original state (i.e. before lifting the partition).

7.3.6. *A Pure De-Assimilation Process*

This section is not essential for understanding either entropy or the Second Law. However, since it is my favorite example, I am taking the liberty of presenting it here. The reader who is not interested in "my favorite" process can skip this section.

We have seen that the assimilation process causes a decrease in *SMI*. This leads us to conclude that the reverse of this process, the *de-assimilation* process, should be associated with an *increase* in the *SMI*. A pure assimilation process is one where the volume does not change, but only N changes. This is simply the reverse of the pure assimilation process in Fig. 7.5.

We already know that a spontaneous process causes an increase in *SMI*, so why do we never observe the reverse of the process in Fig. 7.5 occurring spontaneously? The answer is simple. Many processes can be described for which the change in *SMI* is positive, but they do not actually occur as long as a constraint precludes their occurrence. Indeed, the reversed process as in Fig. 7.5 does not occur spontaneously. However, there exists a pure de-assimilation process which is equivalent to the reversed process in Fig. 7.5, and which occurs spontaneously.

Consider a system of $2N$ molecules, each of which contains one chiralic center — say, an alanine molecule (Fig. 7.7). The two molecules in the figure are mirror images of each other. We initially prepare all the molecules in one form — say, the d form,

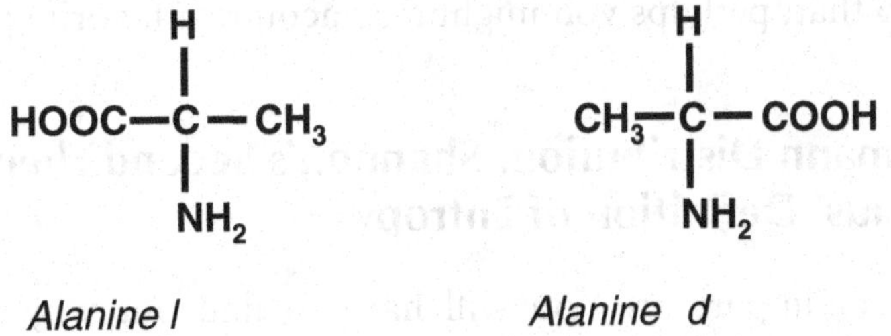

Alanine *l* Alanine *d*

Fig. 7.7 Two mirror images of an alanine molecule.

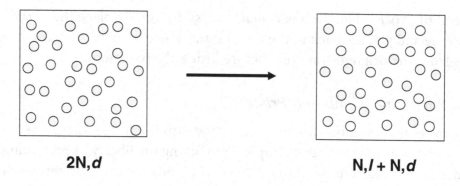

2N,*d*   **N,*l* + N,*d***

Fig. 7.8   A spontaneous *de-assimilation* process.

in a volume V and temperature T. We next introduce a catalyst that facilitates the conversion of d to l form. It is known that this system will spontaneously evolve into a mixture of N molecules in the d form and N molecules in the l form. The net change in SMI in this process is $2N \log_2 2 = 2N$. This process is shown schematically in Fig. 7.8.

This is a *pure* de-assimilation process — the system of $2N$ indistinguishable particles evolved into N particles of one kind and N particles of a second kind. The net effect is the same as in the reverse of the process in Fig. 7.5, where we achieved a split of the $2N$ particles into two groups — N in one box and another N in a second box. In both processes, the only change that took place is the de-assimilation of the particles, therefore the change in SMI is positive.

If you did not follow the story of the process of de-assimilation, you can forget it. It will contribute nothing to understanding the Second Law. However, if you did follow and you want to learn more, I suggest that you also read my book (Ben-Naim (2008)). If you do that, perhaps you might also acquire a favorite process.

7.4. The Boltzmann Distribution: Shannon's Second Theorem and Clausius' Definition of Entropy

This section is very important. You will have to find out why for yourself. We shall first review what we have found experimentally in the games with marbles in Chapter 4. Then we shall translate it into the language of real particles in real boxes.

In doing so, we shall discover the Boltzmann distribution, which is central in statistical thermodynamics. In Section 7.4.2, we shall discuss the particular shape of the entropy versus the energy curve of the system. This will lead us to the identification of the Clausius *definition* of entropy. As a by-product, we shall also find how the temperature is related to the *shape* of the entropy versus energy curve.

In Section 7.4.4, we shall examine how the entropy changes when we transfer energy from one system to the other. This will lead us to Clausius' *formulation* of the Second Law of Thermodynamics.

7.4.1. *The Boltzmann Distribution*

The experiments we did in Chapter 4 were designed to "discover" the Boltzmann distribution, and systems that can exchange energy. The Boltzmann distribution is a very fundamental one in statistical mechanics. Basically, it states that if the particles can occupy different energy levels, and if the *total* energy of the system is fixed, then the particles will occupy the energy levels according to the Boltzmann distribution. This is a particular example known as the exponential distribution in the mathematical theory of probability.

What we have observed is that given a fixed number of particles in a given volume, and given the total energy of the system (the total energy is represented in the game by TL), the system will evolve in such a way that relatively more particles will occupy level zero (or ground zero), and the occupation numbers will drop sharply as the energy levels increase.

The most familiar example is the distribution of molecules in their energy levels; for instance, the vibrational, rotational, or electronic energy levels. Less familiar perhaps is the Barometric distribution. Consider a vertical column of air above sea level (Fig. 7.9). The density of the molecules at each level (i.e. at each height interval) is, under ideal conditions, the Boltzmann distribution. By "ideal conditions" we mean that the molecules do not interact with each other, there is no wind, and the temperature is the same at all heights. These conditions are not met in reality, yet surprisingly, to a good approximation, the density distribution is very close to the Boltzmann one. If one can assume that the gas is ideal (i.e. no interactions between the molecules), then the pressure of the gas at each "level" is proportional to the density at that level.

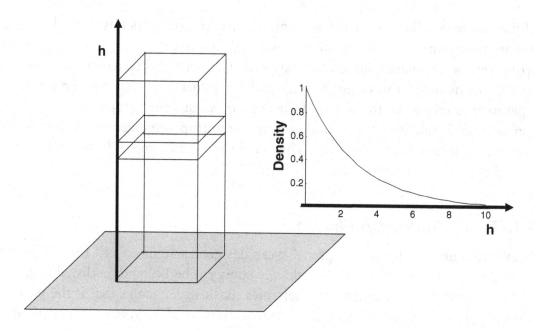

Fig. 7.9   The barometric distribution.

Pilots in older aircrafts used to measure the pressure of the air outside the airplane and "convert" it into "height" or altitude by using the Boltzmann formula.[7]

7.4.2. *Clausius' Definition of Entropy*

In Chapter 4, we found that for any system and any initial configuration, "shaking" the system for some time leads to an equilibrium state. The equilibrium state is characterized by a maximum *SMI*, which is translated here into entropy of the system of real particles. We also found that the equilibrium state is characterized by maximum multiplicity (W). This is important for understanding *why* the system will proceed towards the equilibrium state. The argument is essentially the same as the one discussed in Section 7.2.2 and will not be repeated here.

In Chapter 4, we also found that the *SMI* is a monotonic increasing function of *TL* (Fig. 4.20). Translating *SMI* into entropy, and *TL* into energy, we get the curve of the entropy versus the energy of a system of particles. This is essentially the Sackur–Tetrode relationship between the entropy and the energy of an ideal gas.[8]

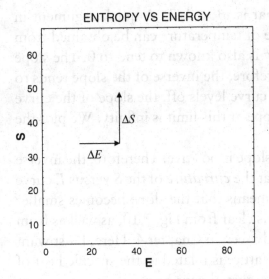

Fig. 7.10 The entropy versus the energy curve.

In Fig. 7.10, we have redrawn the curve of *SMI* versus *TL* from Fig. 4.20, but changed the notations to entropy S versus energy E. The units of entropy and the energy are of no concern to us in the present discussion.

Suppose that we start with a system with some energy E and read from the graph in Fig. 7.10 the corresponding (equilibrium) value of S. We add to the system a *small* quantity of energy, denoted by ΔE. The corresponding increment in the (equilibrium) entropy is denoted by ΔS.

Define the slope of the curve as the ratio of ΔS and ΔE (i.e. $\Delta S/\Delta E$). We can calculate the slope at each point of the curve in Fig. 7.10, we can draw a new curve shown in Fig. 7.11, where we denoted the slope by $\beta = \Delta S/\Delta E$. We shall soon see that the quantity denoted β has all the properties

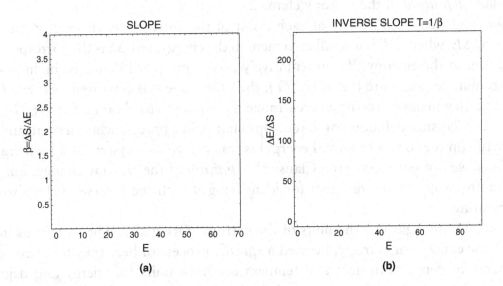

Fig. 7.11 (a) The slopes of the curve in Fig. 7.10. (b) The inverse of the slope.

of the *inverse* of the absolute temperature; that is, $\beta = 1/T$. A quick argument in favor of the identification of β with the inverse of temperature can be obtained from Fig. 7.10. First, at very low energy, the entropy is also known to tend to 0. The slope of the curve in Fig. 7.10 tends to infinity. Therefore, the inverse of the slope tends to 0. At very high energies such that the entropy curve levels off, the slope of the curve in Fig. 7.10 is 0, therefore the inverse of the slope at this limit is infinity. We plot the slope and the inverse of the slope in Fig. 7.11.

At any intermediate point in Fig. 7.10, the slope is positive. Therefore, the inverse of the slope is also positive. Finally, we note that the *curvature* of the S versus E curve in Fig. 7.10 is negative. A negative curvature means that the slope becomes smaller and smaller as we increase the value of E. This is clear from Fig. 7.10, as well as from Fig. 7.11a. This is important, as we have already seen in Chapter 4. Here, I just want to draw your attention to the fact that the curvature is related to the specific heat of the system,[9] and the specific heat of the system must always be positive.

All the arguments provided above are only supportive to my contention that the slope of the entropy versus energy curve is related to the inverse of the temperature. A more convincing argument is given below and in Section 7.4.4.

Here, we shall show that from the shape of the curve in Fig. 7.10 we can recover Clausius' *definition* of the entropy change.

We have defined the slope at each point of the entropy versus energy curve by $\beta = \Delta S / \Delta E$, where ΔE is a small increment in the energy and ΔS is the corresponding increment in the entropy. If you tentatively accept my word that β is the inverse of the absolute temperature (i.e. $\beta = 1/T$), then the slope β is rewritten as $\beta = 1/T = \Delta S / \Delta E$. All you have to do now is rearrange this equation to obtain $\Delta S = \Delta E / T$. This is exactly Clausius' definition of the entropy change in a process where a small amount of energy (in the form of thermal energy) is transferred to a system at a temperature T. Thus, we not only recovered Clausius' *definition* of the entropy change, but also got another supportive argument for identifying β with the inverse of the absolute temperature.

A final note regarding the units of the entropy: Originally, when Clausius introduced the concept of entropy, he used a *specific* process of heat transfer at constant temperature. Since both heat and temperature have units (of energy and degrees, respectively), the entropy change got the units of energy divided by Kelvin (the units of the absolute temperature). This is a historical accident. Had the kinetic theory of gases preceded Clausius, the temperature would have been given the more natural units of energy. This would render the entropy a unit-less quantity. It would also have eliminated much of the confusion as well as the mystery about the meaning of the entropy.

7.4.3. *Energy Transfer Between Two Systems in Contact*

In Chapter 4, we carried out two processes of transfer of energy between two identical systems (i.e. with the same number of particles and the same volumes), but with different energies. We then transferred a given quantity of energy (represented by TL) from the high energy system to the low energy system. We found, as we expected from the known dependence of the entropy on the energy (Section 7.4.1), that the entropy of one system increased while the entropy of the second system decreased. This could be read from the plot of entropy versus energy (or *SMI* versus TL). We also found that the *net change in entropy of the combined systems has increased*. This is almost the original Clausius formulation of the Second Law:

> For two identical systems (i.e. having the same number of particles N, and the same volume), but having different energies, say $E_2 > E_1$, if we transfer a small amount of energy from the E_2 system to the E_1 system, the total entropy of the combined system will increase.

I said this is "almost" the Clausius *formulation* of the Second Law, and not *the* Clausius formulation, because we still do not have the concept of temperature, and we have restricted ourselves to *equal* systems and not to any two systems. Nevertheless, the conclusion we have reached here is important. It follows from the fact that entropy is not only monotonically increasing function of the energy but it also has a negative curvature (i.e. the slope of the function decreases as we increase the energy). We shall discuss the Clausius formulation of the Second Law further in the next subsection.

7.4.4. *Spontaneous Flow of Energy from One System to Another*

Perhaps the most important experiment carried out in Chapter 4 was the spontaneous "flow" of energy (represented by TL) from the high-energy system to the low-energy system. We first restricted ourselves to two *identical* systems, except for having different amounts of energy.

We brought the two systems into contact in the sense that they could exchange energy (keeping the number of particles and the volume of each system unchanged). We found that energy flowed spontaneously from the high energy system to the low energy system. In this process, the net change in entropy was positive. The flow of energy continued until the energies of the two systems were equal. If we start with E_1 and E_2, where $E_2 > E_1$, then at the final equilibrium the energy of each system will be $(E_2 + E_1)/2$. At this state, the entropy as well as the probability of the equilibrium state of the combined systems will be maximal.

At this point, we could formulate another version of the Second Law. However, to make the statement more general and to bring the formulation closer to *Clausius' formulation*, we need to remove the restriction of the two identical systems.

Suppose that we start with two unequal systems — say, one with N_1 particles in volume V and the second with N_2 particles in volume V, and $N_2 > N_1$. The graphs of the entropy as a function of energy for these two systems have roughly the same shape, but the values of the entropies are higher for the larger system. Look at Fig. 4.27, and translate *SMI* into entropy and *TL* into energy.

We now bring the two systems to thermal contact in the sense that they can exchange energy but not particles. What will happen is that energy will flow from one system to the other system. As a result of this flow, the net entropy of the combined system will increase. However, in contrast to the case of the two identical systems, in the present experiment the flow of the energy will not stop when the *energies* of the two systems are equal, but rather when the *slopes* of the curves of the entropy versus energy will be equal.

Here is the sweet raisin that was hidden in the cake. We know from thermodynamics, as well as from everyday experience, that when two different bodies are brought into thermal contact, heat will flow from the system of higher *temperature* to the system of lower temperature, until the combined system reaches a uniform temperature, which is somewhere between the two initial temperatures of the two systems.

In our experiment in Section 4.8, we found out how much energy (*TL*) was transformed from the "bigger" to the "smaller" system. We also found that at the equilibrium point, when the entropy (*SMI*) and the $\log_2 W$ of the combined system has a maximum, the slopes of the two curves of the entropy versus energy of the two systems were equal. We also found that energy flowed from the system which initially had a *smaller* slope to the system having the *larger* slope. The last statement, when translated in terms of the *inverse* of the slope, is nothing but Clausius' *formulation* of the Second Law:

> At thermal contact between two systems, energy (heat) will always flow from the body having higher temperature to the body having the lower temperature. In this process, the entropy of the combined system will always increase, until it reaches a maximum. At this point the flow of energy will stop. The combined system will reach a new equilibrium state at which the temperatures of the two systems are equal.

Why does the energy flow from one system to another? Because the probability of the new equilibrium state of the *combined* system is larger (much larger!) relative to the probability of initial state of the combined system.

Note carefully that here we have discussed Clausius' *formulation* of the Second Law, whereas in Section 7.4.3 we recovered Clausius' *definition* of the change in entropy in a *specific* process of heat transfer. Other spontaneous processes such as expansion or mixing of gases have different entropy changes, involving neither energy transfer nor temperature. Thus, Clausius' *definition* applies to a *specific process*. So is Clausius' *formulation* of the Second Law, which also applies to a specific process.

The general formulation of the Second Law is:

> For any spontaneous process in an isolated system, the entropy always increases until the system reaches an equilibrium state, at which state the entropy is maximum.

7.5. The Maxwell–Boltzmann Distribution: Shannon's Third Theorem and Clausius' Formulation of the Second Law of Thermodynamics

In Section 7.2, I have claimed that understanding the expansion process ranks second in importance to understanding the Second Law. This section is devoted to a process which is the *first* in its importance to understanding the Second Law.

This section will be shorter than Section 7.4, but no less important. Shorter, because all that is written is Section 7.4 applies to the process we shall discuss in this section. Therefore, we shall discuss only the new aspects of the system that we have studied in Chapter 5, namely systems which are characterized by the

Maxwell–Boltzmann distribution. In Section 7.5.1 we shall translate from the marble language of Chapter 5 (and Chapter 6) into the language of real particles. In Section 7.5.2, we shall briefly recapitulate what we have discussed in Section 7.4 and apply it to the *MB*-case. Finally, in Section 7.5.3 we shall discuss some new aspects of the *MB*-case.

7.5.1. *Translation*

In the expansion process, the cell's numbers represent the *location* of the particle. In the present case, the cell's number represents the *velocity* of the particles. The string length, TL, in Chapter 4 represented the energy of the particles. In Chapter 4, as well as in Section 7.4, we did not specify what kind of energy the particle possess. In Chapter 5, the cloth area, TA, represented the *kinetic energy* of the particles. The velocity of the particles can be either positive (running forward) or negative (running backwards). In the real world, one usually discusses the absolute value of the velocities, sometimes referred to as the *speed* of the particles, which in one dimension is simply the absolute value of the velocity. For the present book and for understanding entropy and the Second Law, we do not need the three-dimensional case or the absolute velocities of the particles.

We have a system of N particles in a one dimensional "box" of length L.[10] The molecules can fly either in the positive direction, with velocity v, or in the negative direction with velocity $-v$. Which direction is chosen as the positive or negative is of no importance; we can choose for instance the direction from left to right to be the "positive" direction, as the direction in which you write English. But, if you prefer, you can choose the right-to-left direction, as in writing Hebrew, as the positive direction.

The important quantity is not the velocity itself but the *square* of the velocity, v^2. It turns out that the kinetic energy of the moving particle is proportional to the *square* of the velocity. Therefore, it does not matter in which direction the particles move; the kinetic, or motional energy, of the particles is proportional to v^2.

Thus, a bullet impacting a body causes an effect which is proportional to v^2, not to v itself. Once we recognize that v^2 is related to the energy of the particle, we can apply all that we have said in Section 7.4 on the energy distribution, the entropy versus energy curves, and so on, to the energies of the moving particles. We have only to switch from the TL language to the TA language;[11] both represent the energy of the particles. The only difference is that now we discuss the kinetic energy of the moving particles explicitly.

One caveat before we continue: In the expansion process, we described the locations of the particle and the changes in the locational distribution of the particles (the cell's number in the case of marbles or the location x of a particle in space). We also noted that the "shaking" of the system of real particles is achieved by the *motions* of the molecules. In this section, we shall describe the distribution of *velocities* of the particles. In this case, the motion of the particles plays two roles: as the objects of the distribution (i.e. how many particles have such and such velocities) and as the factor underlying the "shaking" mechanism which induces changes in the distribution of the velocities.

7.5.2. *Summary of What We Already Know*

All we have said in Section 7.4 may be applied to the *MB*-case. We only have to replace *TL* with *TA*, and *TA* with *kinetic* energy of the particles. In fact, once we know that the kinetic energy of the particles is proportional to the square of the velocity, we could even "predict" the Maxwell–Boltzmann distribution, simply by changing variables from the energy to square velocity.[12] Also, the curves of the entropy versus energy (which we translated from the *SMI* versus *TL* curves) can be applied to this case without any change. The only change is in the language we use. Instead of saying that the particles have energy *E*, we shall say that the particles have *kinetic energy E*, and this energy is proportional to the average of the square of the velocity. Instead of having a system with constant energy (translated from *TL* constant), we shall say that the total kinetic energy is constant (translated from *TA*).

Maxwell was the first to show that the distribution of velocities (here in one dimension) is the Normal distribution. Boltzmann has shown that a system of particles moving with any initial distribution of velocities will reach an equilibrium distribution which is the Normal distribution. This Normal distribution is extremely important in statistics and probability theory. When applied to the velocities of particles, it is called the Mawxell–Boltzmann distribution.

In the case of marbles in a box, we limited the *locations* of each marble to a fixed number of cells. For real particles, we must limit the locations of the particles within a box of volume V (the box could be of one, two, or three dimensions). It is important to limit the range of variation of the locations to a finite box. If one takes a box with any number of particles and opens it so that each particle is not constrained to move within given boundaries, then after a short time all the particles will fly with a constant velocity and direction indefinitely. Such a system will not reach an equilibrium state.

In contrast, the constraint on the system discussed in Chapter 5 was on the total *kinetic energy* of the particles. There was no constraint imposed on the range of the velocities. The particles can, in principle, have any velocity between minus infinity and plus infinity.

However, in practice, fixing the total kinetic energy of the particles leads to practical limits on the possible range of velocities. Furthermore, fixing the total kinetic energy *determines* the distribution of velocities. This is what Shannon has proved, and this is what we have discovered experimentally in Chapter 5. Translating from Chapter 5, we can say that if we fix the total kinetic energy of the system (TA constant), the system will always proceed to an equilibrium state. This state is characterized by a velocity distribution of the *MB*-case. It is also characterized by a maximum entropy and maximum probability.

We could have concluded this chapter with the statement made in the beginning of this section; namely, go to Section 7.4 and replace TL with TA and TA with E, and E with the total kinetic energy. However, because of its historical importance, we shall discuss one process where we can follow how the *MB* distribution is changing upon thermal contact between two bodies.

7.5.3. *Spontaneous Heat Transfer from a Hotter to a Colder Body*

In Section 7.2, we analyzed the molecular reasons for the spontaneous expansion of particles from a smaller volume to a larger volume. With that process we could formulate the Second Law of Thermodynamics, a formulation which is equivalent to any other formulation of the Second Law. Note carefully that we did not invoke the Second Law to explain the process of expansion. We rationalized the process of expansion, and then we formulated the Second Law.

Historically, the simplest formulation of the Second Law was Clausius' formulation: Heat always flows spontaneously from a hotter to a colder body (assuming that the system comprising the two bodies is isolated).

Clausius' formulation is what every one of us already knows. Heat always flows in one direction. We never see a cold body becoming colder by transferring heat to a hotter body which becomes hotter.

But why? Can we rationalize this process without invoking the Second Law?

That is exactly what we shall be doing in this section. In this case, the rationalization is not as straightforward as it was in the case of the expansion process that we discussed in Section 7.2.2. However, we have all the necessary facts and tools to undertake this task.

Suppose we have two systems. For simplicity we assume that each contains N atoms, say of argon, in a volume V. The two systems have different total kinetic energies (this is the analog of TA). Let us assume that system A has energy E_A and system B has energy E_B, and that $E_B > E_A$. We already know from the experiments carried out in Chapter 5 that if we bring the two systems to thermal contact (this simply means that we let the two systems evolve towards equilibrium keeping the *sum* of the energies constant), heat (here, kinetic energy) will flow from the high energy system to the low energy system. Now, we want to look into the molecular mechanism underlying this process. The simplest way to do this is to open a small window between the two systems for a short time period in such a way that a small number of particles will be exchanged between the two systems. For the moment, it does not matter in which direction the particles will move. In a short time when the window is open, some N_A, A particles will flow from system A to system B and N_B, B particles will flow from system B to system A. For simplicity, we assume that N_A and N_B are nearly the same number of particles. Based on what we already know, system A, with the smaller average kinetic energy, has a distribution of velocities sharper than the distribution of the velocities of system B, the one with a lower average kinetic energy.

The N_A, A particles which originated from system A have a sharper distribution of velocities than the molecules in B, whereas the N_B, B particles originating from B have a flatter distribution of velocities than the particles in system A.

Based on the experiments carried out in Section 5.6, we can conclude that the distribution of velocities in system A will become a little flatter and the distribution in system B will become a little sharper (see Figs. 5.19 and 5.21).

Clearly, if we repeat this process and let more particles exchange places, the overall distribution of velocities in A will become flatter and the distribution of velocities in B will become sharper. This process will stop when the distribution of velocities in the two systems becomes equal (as in Figs. 5.20 and 5.21).

Note that we have achieved this equalization of the distribution of velocities by transferring of particles which carry with them different energies. At the final equilibrium state the average number of particles in each system has not changed. Therefore, only thermal energy has been exchanged between the two systems. We could have achieved the same final equilibrium by introducing a heat-conducting material between the two systems. In this case, however, the description of the molecular process will be more complicated.

To summarize, when we have two systems having different velocity distributions at contact, the combined system will always evolve towards a new equilibrium system. The distribution of one system will gradually become flatter and the distribution of the second will become sharper. At equilibrium, the two systems will have the same distribution of velocities.

This result is essentially equivalent to Shannon's third theorem, which in our example states that under the constraint of constant kinetic energy, the equilibrium distribution of velocities (in one dimension) will be Normal; that is, the Maxwell–Boltzmann distribution. We also know that this process will occur towards that end because the probability of the Normal distribution is maximal over all other possible distributions.

7.6. The Conclusion of All Conclusions

To understand *what* entropy is and why it always changes in one direction (the Second Law), it is enough to analyze two fundamental processes: the expansion of a gas and the heat transfer from a hotter to a colder body.[13]

Of course, there are myriads of other processes that are governed by the Second Law. Some of these are simple extensions or generalizations of the processes we have studied in this book. Some others are too complicated to be studied on a molecular level, such as processes of life and death. Yet, we believe that in all the spontaneously occurring processes, entropy increases.

We have studied several processes involving marbles in cells. We have also analyzed several fundamental real processes: expansion of gas and heat transfer.

It is time to summarize the elements that are common to all these processes, and by generalization, to all the spontaneous processes occurring in nature. These common features are encapsulated in the concepts of entropy and the Second Law.

On the most elementary level we have a *specific* arrangement, or a specific state of the system (the locations and velocities of each *specific* particle[14]). However, since particles are indistinguishable, we cannot follow the evolution of the *specific* state of the system (to where each specific particle goes, and how the velocity and direction of each particle changes). All we can follow is the dim-state of the system (how many particles are in this or that location and have velocities of this or that magnitude). In classical physics, the locations and the velocities are conceived as being continuous variables. However, in practice, we cannot determine the exact values of either the locations or the velocities of the particles.[15] Therefore, it is both convenient and practical to envisage a finite number of cells in which a particle may be, and a finite number of bins, each representing a small range of velocities. This is similar to what we have done in describing the system of marbles in cells.

Once we have agreed about the cell division of the entire space of locations and velocities (here in one dimension as depicted in Fig. 7.12, but it can easily be extended to three-dimensional space of locations and velocities), then the dim-state may be described by a distribution of particles — say, $N(x, v)$ particles having location x and velocity v. This gives us a full dim-description of the state of the system. From the numbers $N(x, v)$ we can construct the probabilities $p(x, v) = N(x, v)/N$. Each of these quantities are related to the probability of finding a particle (any particle) in location x and having velocity v. This is the analog of the distribution of marbles in cells. We can write the distribution as $(p_1, \ldots, p_n)$ where p_i is the probability of finding a marble in cell i, or the probability of finding a particle in state i. The

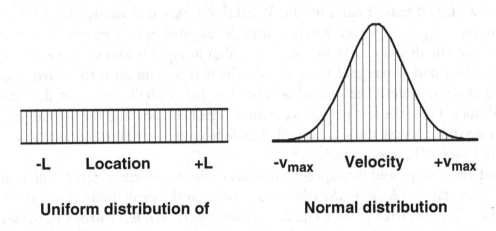

$-L$ **Location** $+L$ $-v_{max}$ **Velocity** $+v_{max}$

Uniform distribution of   **Normal distribution**

Fig. 7.12   (a) Discretization of the range of locations between $-L$ to $+L$. (b) Discretization of the range of velocities, practically from $-v_{max}$ to $+v_{max}$.

state i could be a location, a velocity, or some internal energy level of the molecule. The important thing is that we describe the dim-state of the system by a *dim-state probability distribution.*

On this *dim-state probability distribution* one can define many quantities, such as the *SMI* and *W*. Remember *SMI* is a measure of the *size* of the missing information associated with the probability distribution. The quantity *W* counts the number of specific states "belonging" to the dim-state.

As the system evolves with time, the dim-state also changes. In the games of marbles, we have followed the distribution of marbles in the different cells. In a real system, we might follow the distribution of particles in different locations or velocities.

Since the quantities *SMI* and *W* are *defined* for each dim-state, the evolution of the dim-states induces an evolution of the *SMI* and of *W*. These were the two quantities that we have chosen to follow in all our experiments with marbles in Chapters 3–5. In a real system, with real particles, the *SMI* has the same meaning as the *SMI* in the game of marbles, and *W* is related to the *probability of the dim-state.*

Let us call the latter probability the *super* probability, denoted *PR*. This should be distinguished from the probabilities p_i that comprise the *dim-state probability distribution* $(p_1, \ldots, p_n)$. p_i is the probability of finding a particle in some location, or velocity (or cell) denoted i. On the other hand, the probability *PR* is defined on the *entire* distribution $(p_1, \ldots, p_n)$, so does the quantity *SMI*.

When we conducted the experiments with marbles, we followed the evolution of *W* and we found that at equilibrium, *W* attains a maximal value. The same is true for the evolving real system. Whether it is the expansion of a gas or heat transfer, the system will always evolve in such a way that at equilibrium we have a maximal value of *SMI* and a maximal value of *W*. The first maximum is the *entropy* of the system at equilibrium. The second is related to the probability of the dim-state at equilibrium. Thus, the first provides the *meaning* of entropy and the second provides an explanation of why the system will change from the initial to the final states (i.e. Second Law of Thermodynamics).

Both the entropy and the super-probability of the equilibrium state are determined by the *dim-state probability distribution.* (Note that the probability distribution can be either UDG or NUDG — see Fig. 2.19.) One describes *what* entropy is, the second *why* it changes in a one-way direction. These two aspects of the Second Law seem

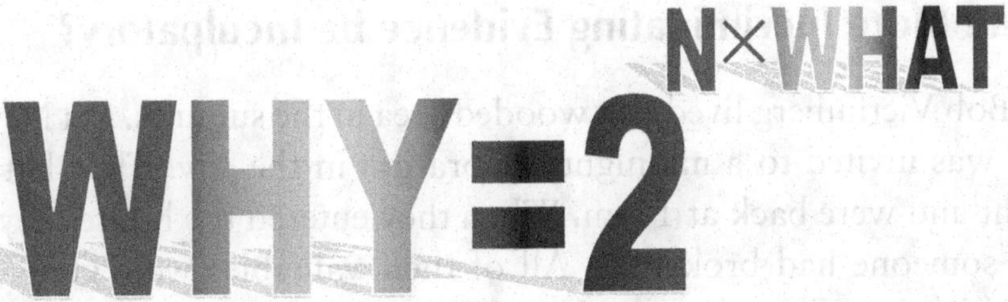

WHY=2 N×WHAT

Fig. 7.13 The symbolic relationship between the questions of *what* and *why*.

to be unrelated, and yet they are closely related to each other. This relationship is shown symbolically in Fig. 7.13. An artist's view of this relationship is shown in Fig. 7.14 (as well as in the cover design of this book). If you are curious to see what the mathematical relationship looks like, see Note 16.

Fig. 7.14 Artist's view of the relationship between entropy and the super probability (drawn by Alexander Vaisman).

Snack: Can More Incriminating Evidence be Inculpatory?

Mary and Bob Victimberg lived in a wooded area in the suburbs. On New Year's Eve, the couple was invited to a midnight celebration in the city. They left the house at 12 midnight and were back at 1 a.m. When they entered the house they immediately knew that someone had broken in. All of their valuables were gone. The infrared detector, which is activated every time there is some movement in the house, was blinking. They checked the detector and found that for exactly 30 minutes, someone had been moving in the house. The reading on the metal detector, however, did not tell them at exactly what time the movement had been detected. All that they could say was that it happened between 12 midnight and 1 a.m., the timeframe during which they were out of the house. They immediately called the police to report the break-in. They were asked if they had any suspects in mind but they said they could not say anything. The only thing they could say with certainty was that they were out of the house between 12 midnight and 1 a.m., and that the infrared detector was activated for half an hour between 12 midnight and 1 a.m.

The following day, Jerry Pick, a well known criminal, was caught for shoplifting in the supermarket not far from where the Victimbergs lived. While the county policeman was writing down the case in the police blotter, the thought crossed his mind that Pick must have been involved in the Victimberg's household break-in, so he decided to detain Pick for the night. Pick denied any involvement in the break-in.

Pick's detention made it into the local papers as well as onto the television, with his face plastered on the TV and in the papers. News features both on TV and in the papers mentioned that there had been an unsolved robbery the day before, and that the police requested the public to come forward if they had seen any suspects in the vicinity of the Victimberg's house on New Year's Eve between 12 midnight and 1 a.m. It was likewise mentioned in the news that the burglary took place for exactly 30 minutes, as the infrared reading had indicated, although the exact time of the break-in was not known.

Chief Officer Greenhead was sitting in a bar nursing a mug of beer when the TV screen flashed Jerry's face. The bar was a mere five-minute walk from where the Victimbergs lived. Chief Greenhead had all eyes and ears on the TV news about the two robberies of the two consecutive days when it struck him like lightning as he remembered that Jerry had entered the very same bar on New Year's Eve at exactly 12.35 a.m. He also vividly recalled that Jerry had stayed in the bar for about an hour. Doing a mental calculation and after considering Jerry's criminal record, Chief Greenhead was convinced that Jerry could have committed the robbery between 12 midnight and 12.30, and had enough time (five minutes) to get from the Victimberg's residence to the bar. With this sure evidence he wrote down all of his recollections from New Year's Eve and sent them to the court, fully convinced that his circumstantial evidence would bolster the evidence against Jerry in his involvement in the Victimberg's robbery.

At exactly the same time that Chief Greenhead was rushing to file his report in the bar, Officer Redhead was dining in a restaurant nearby. Officer Redhead was just as familiar with Pick's criminal record and, by coincidence, he had also seen Jerry dining in the same restaurant where he was on the night of the alleged break-in at the Victimberg's. He vividly recalled that Jerry had dined there at 11 p.m. — just before midnight — and had left the restaurant at 12.25 a.m. Watching the news on TV, he, like Chief Greenhead made a quick mental calculation and concluded that Jerry had had enough time to get from the restaurant to the Victimberg's residence after he had left the restaurant, stayed and robbed the house in 30 minutes, and then sped off. With hands trembling, he wrote down all the details and was satisfied with his analysis. With a glimmer of hope, he felt confident that his accurate details and analysis would land him a promotion.

Almost simultaneously, the two reports were well on their way to court, with bearers who were equally confident of the veracity of their details. Both believed that their reports contained highly plausible evidence which could pin down Jerry regarding his involvement in the New Year's Eve break-in. Jerry, of course, denied the allegations.

The next morning, the two reports were laid out on Judge Judy's desk. She first read Chief Greenhead's report. It was very convincing, considering Pick's criminal records. So convincing was Chief Greenhead's report that Judge Judy almost issued an extension of Pick's detention in the county jail. Upon reading Officer Redhead's report, however, which also contained very convincing evidence against Pick, Judge Judy decided to let Pick off the hook.

Can you explain why Judge Judy changed her mind?

Answer: Initially, the judge could have reached some conclusion regarding the probability that the suspect is guilty. Let us say that she assumed that the probability of the suspect's guilt given his past criminal record was 50%, and we write it in shorthand notation as:

$$Pr \text{ (guilt/past record)} \approx 50\%$$

After reading the evidence presented by Chief Greenhead, the judge could modify her estimate upwards:

$$Pr \text{ (guilt/past record } and \text{ Greenhead evidence)} \approx 80\%$$
$$> Pr \text{ (guilt/past record)} \approx 50\%$$

The same conclusion could be reached from Redhead's evidence:

$$Pr \text{ (guilt/past record } and \text{ Redhead evidence)} \approx 80\%$$
$$> Pr \text{ (guilt/past record)} \approx 50\%$$

We can say that the evidence of each of the officers reckoned *by itself* supports the contention of the defender's guilt. However, *taken together* the two pieces of evidence exonerate the suspect:

$$Pr \text{ (guilt/past record } and \text{ Greenhead } and \text{ Redhead evidence)} \approx 0$$

Clearly, it is known that if the suspect had spent about an hour in the restaurant and left the restaurant at 12.25, and if it is also known that he entered the bar at 12.35 and stayed there for an hour, he could not possibly have spent the full 30 minutes in the Victimberg's residence. Thus, although *each* of the pieces of evidence is supportive, the combined evidence is unsupportive. Venn's diagrams relevant to the story are shown in the figure.

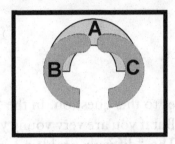

B Supports A and C Supports A.
B AND C is an Impossible Event.

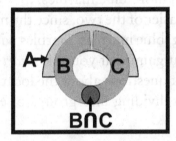

B Supports A and C Supports A, but
B AND C Makes A an Impossible Event.

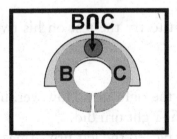

B Supports A and C Supports A, but
B AND C Makes A a certain Event.

Venn diagrams

CHAPTER 8

Notes

Notes to Preface

1. See Section 2.2 for an answer to this question. In the "easy problem" you should choose the urn on the right. But if you are very young you might choose the correct urn for the wrong reason. The "difficult problem" is more difficult because the correct choice is the urn on the left, but young children might choose the urn on the right because it contains more blue marbles.
2. The "easy" problem is the easier of the two, since the marbles are already arranged in groups of equivalence (e.g. blue marbles, marbles with two colors, real marbles).
3. Yes, you can guarantee your gaining if you ask *smart* questions (see Section 2.3).
4. If you are not allowed to ask questions about the locations of the marbles, you can still ask *smart* questions by dividing into groups of equivalence. See Section 2.3 for more details.

Notes to Self-Testing Kit

I asked my friend John Knight to try the test on his two young sons. Here are the results:

Michael — Age 10
Easy problem — Mike picked the right urn. However, he picked that urn because it had more blue marbles than the right urn did.

Difficult problem — Mike picked the left urn. Amazingly, he picked the left urn because the right urn had more red marbles whereas the left urn had the same number of each color.

Brandon — Age 7

Easy problem — Brandon picked the left urn because the right urn had more red marbles than the left urn. Clearly, Brandon was looking at the problem from the perspective of avoiding the red marbles.

Difficult problem — Brandon picked the left urn. Again, his reasoning was that the left urn had less red marbles than the right urn.

Regarding the Measure of Information Test:

Each took the test separately. Here are the results:

Brandon - Age 7
Easy Problem

1. Is it real? No.
2. Does it have a dot? Yes.
3. Does it have a red dot? No.
4. Is the marble blue? No.
5. Is it red (with a dot)? Yes.

Difficult problem

1. Is it real? Yes.
2. Does it have a dot? Yes.
3. Is it green? No.
4. Is it orange? Yes.

Brandon started off with an apparent strategy but, by the third question, he was randomly picking marbles.

Questions:

1. Why were the problems labeled easy and difficult? Because on the difficult one they are all mixed up.
2. Can you guarantee you will again? No, not every time. Question 3 did not really apply since he did not ask any questions dependent on location.

Michael — Age 10
Easy problem

1. Is it real? Yes
2. Does it have a dot? Yes
3. Is it a cool color? No
4. Is it orange? No

Difficult problem

1. Is it a computer graphic marble? No
2. Does it have a dot? No
3. Is it a hot color? No
4. Is it blue? Yes

Questions:

1. Why were the problems labeled easy and difficult? Because in the easy one the marbles are in straight rows and organized. They are mixed up and not organized in the difficult problem.
2. Can you guarantee you will again? No. Why not? Because I am not that good at math and I'm not psychic.
3. Again, question 3 did not apply.

The two boys were asked to play the 20Q game with Figures 2.4 and 2.14.

   Both boys asked either specific questions (is it the dolphin?), or selected a group of objects according to some attribute (is it blue?). Both have more difficulty with Figure 2.14 because they did not identify the persons in the figure. It did not occur to them that they could have asked smart questions without identifying the persons (e.g. is the person in the left hand half of the figure?)

Notes to Chapter 1

1. The efficiency of a Carnot engine is simply given by

$$efficiency = \frac{T_2 - T_1}{T_2} = 1 - \frac{T_1}{T_2} \leq 1 \tag{1}$$

   where T_1 and T_2 are the lower and the upper temperatures between which the heat engine operates.

2. The entropy is said to be a "state function." What that means is that given the thermodynamic *state* of a system — say, by describing its volume, energy, and composition — the entropy of the system is *determined* by these parameters. Clearly, when a system moves from one state, A, to another state, B, the change in the state function depends only on the states A and B, and not on the path along which the system has moved from A to B.
3. The definition of entropy change in thermodynamics involves a small amount of heat transferred dQ at a given temperature T:

$$dS = \frac{dQ}{T} \tag{1}$$

4. Quoted by Cooper (1968).
5. Cooper (1968).
6. The Boltzmann definition of entropy is

$$S = k_B \log W \tag{2}$$

where W is the number of states, or the number of possible arrangements of a given system, and k_B is the so-called Boltzmann constant.
7. Shannon's measure of information is defined on any probability distribution $p_1, \ldots, p_n$

$$SMI = -\sum_{i=1}^{n} p_i \log p_i \tag{3}$$

8. Tribus and McIrvine (1971).
9. Lewis (1930).
10. Boltzmann (1964).
11. Ben-Naim (2008)

Notes to Chapter 2

1. In case (i), the probability of gaining the specific prize in the Indian currency is 1/12 in lottery (a), but it is 1/18 in lottery (b). Therefore, I shall choose lottery (a) if I am interested specifically in the Indian currency. In case (ii), the probability

of getting the maximal prize (which is 10 US$) in lottery (a) is 1/2, but 2/3 in lottery (b). Therefore, I shall choose lottery (b) in this case.

2. Olver and Hornsby (1966).

3. Steven Pinker, in his recent book *The Stuff of Thought* (2007), opens with the tragic events of 9/11:

> A hijacked airplane crashed into the World Trade Center's north tower at 8.46 a.m. Just a few minutes later, at 9.03 a.m. a second plane crashed into the south tower. Thereafter, as a result of the raging fires, both towers collapsed. That tragic incident, as we all know, was perpetrated by Osama bin Laden, leader of the Al Qaeda terrorist organization.

The question Pinker discusses is that of how many events took place in New York on that tragic morning. One could argue that there was only one event. The attacks on both buildings were part of a single plan hatched by a single man ruled by one agenda. Others could argue that there were two events. The north tower and the south tower were two distinct structures, which were hit at different times and went out of existence at different times. These two different views were not a question of *mere* semantics. These opposing views were the focus of the legal dispute between Larry Silverstein, the World Trade Center's leaseholder, and the insurance company. Silverstein held insurance policies which stipulated a maximum reimbursement for each destructive "event." If 9/11 comprised a single event, Silverstein was entitled to receive only 3.5 billion US$, whereas if what happened was considered to comprise two events, he stood to receive 7 billion US$. The lawyers for the leaseholder defined the collapses of the World Trade Center in physical terms (e.g. two collapses), whereas the insurance companies defined it in mental terms (e.g. one plot). You see, there is nothing "mere" about semantics!

4. Piaget and Inhelder (1951).

5. For a review see Fischbein (1975).

6. In the experiments reported by Falk, Falk, and Levin (1980), they also used pairs of roulettes and pairs of tops, each with two colors.

7. See simulated games at my site: http://www.ariehbennaim.com.

8. Shannon's measure of information is defined for any distribution $(p_1, \ldots, p_n)$ by

$$SMI = -\sum_{i=1}^{n} p_i \log p_i \qquad (1)$$

If the distribution is uniform; i.e. $p_i = 1/n$, then

$$SMI = -\sum \frac{1}{n} \log \frac{1}{n} = \log n \qquad (2)$$

For the case that there are two possibilities

$$SMI = -p_1 \log p_1 - (1 - p_1) \log p_1 \qquad (3)$$

This function has a maximum when $p_1 = 1/2$. If we also use logarithm to the base 2 we get

$$SMI = -\frac{1}{2} \log_2 \frac{1}{2} - \frac{1}{2} \log_2 \frac{1}{2} = 1 \qquad (4)$$

This is the maximum value of the *SMI* for the case of two outcomes.

9. If the number of boxes is of the form 2^n, where n is an integer, then the *ANOQONTAITSS* is exactly n. But in the more general case, you need, on average, $\log_2 NB$ questions.

10. This can be calculated from Shannon's measure, knowing the distribution of the letters in the English language. The result is $SMI = 4.14$. See Ben-Naim (2008).

11. Answers to the exercises E2.3 and 2.4.

In the first case, I shall need four questions. The first question will tell me on which board the dart hit. Then I need three more questions to determine the specific square within the board. In the second case, I shall need three questions to find out where the dart hit one board, and another three questions for the second dart. In this case, the number of questions is $6 = 3 + 3$.

12. The *SMI* in this case is:

$$-\left[\frac{1}{2} \log_2 \frac{1}{2} + \frac{1}{4} \log_2 \frac{1}{4} + \frac{1}{8} \log_2 \frac{1}{8} + \frac{1}{16} \log_2 \frac{1}{16} + \frac{4}{64} \log_2 \frac{1}{64} \right] = 2$$

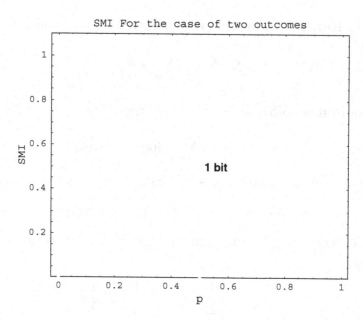

Fig. 2.27 The Shannon measure of information for the case of two outcomes with probabilities p and $1 - p$.

13. The *SMI* for this case is:

$$-\left[\frac{7}{64}\log_2\frac{1}{64} + \frac{57}{64}\log_2\frac{57}{64}\right] \approx 0.8$$

14. The SMI for exercise E2.5 is:

$$-\left[\frac{999}{1000}\log_2\frac{999}{1000} + \frac{999}{999000}\log_2\frac{1}{999000}\right] \approx 0.02$$

15. Mosher and Hornsby (1966).
16. Siegler (1977).
17. Marschark and Everhart (1999).

Notes to Chapter 3

1. Here, we use the so-called classical definition of probability. This is the ratio of the number of specific states belonging to a dim-state, W, and the total number

of specific states W_T.

$$\text{Pr (of a dim-state)} = \frac{W(\text{of the dim-state})}{W_T}$$

2. The *SMI* is defined as $-\sum p_i \log_2 p_i$, where p_i is the probability of finding a specific marble is cell i. If the cells are of equal probabilities $p_i = 1/2$ in this case, then the *SMI* is simply $-\sum p_i \log_2 p_i = \log_2 2 = 1$.

3. The number of specific states belonging to the dim state $(n, NM - n)$ is

$$W = \binom{NM}{n} = \frac{NM!}{n!(NM - n)!}$$

The total number of specific states is 2^{NM}. The probability of the dim state is thus

$$PR(n, NM - n) = \frac{\binom{NM}{n}}{2^{NM}}$$

Notes to Chapter 4

1. The theoretical Boltzmann distribution for the case discussed in Section 4.1 is obtained by maximizing the Shannon function

$$SMI = -\sum_{i=1}^{3} p_i \log p_i \tag{1}$$

Subject to the two conditions

$$\sum_{i=1}^{3} p_i = 1, \quad \sum_{i=1}^{3} (i - 1) p_i = 2/3 \tag{2}$$

This gives the solution

$$p_i = 0.867(0.593)^i \tag{3}$$

Hence the distribution that maximizes *SMI* is

$$p_i = 0.514, \quad p_2 = 0.305, \quad p_3 = 0.18 \tag{4}$$

Which satisfies the two equalities (2). To obtain the limiting average number of marbles in each box for the game in Section 4.1 we have to multiply the results in (4) by 6, to obtain the values of 3.085, 1.829, and 1.081. These are drawn as black points in Fig. 4.6b. The corresponding *SMI* for the theoretical result is

$$SMI = -\sum p_i \log p_i = 1.462$$

2. Answer to Exercise E4.1

The total number of specific configurations is simple to calculate but it is not easy to write them all. The first marble can be placed in one of the three cells. The second marble can be placed in one of three cells, and so on for each of the 12 marbles. Therefore, the total number of specific arrangements is

$$3^{12} = 53,1441 \tag{1}$$

Next, the total number of dim-configurations is more difficult to calculate. You can write some of these; altogether, there are 91 different dim-configurations. However, when you restrict your choice for only those dim-configurations having fixed total length, the number is drastically reduced. For case (i), $TL = 2$, there are only two possibilities; for case (ii), $TL = 5$, there are three; for case (iii), there are seven; and for case (iv) three; and for case (v) there are two.

Note the symmetry about case (iii) for which the limiting distribution is the uniform one. Figure 4.18 summarizes the results for the limiting distribution in each case.

3. The slope is related to the inverse of the absolute temperature, and the curvature is related to the heat capacity of the system. See also Chapter 7.

4. The challenging exercise. Exercise E4.4.

Once you discover that the inverse slope is the criterion for the spontaneous flow of string from one system to the other, examine Fig. 4.27, you draw a graph of the inverse slopes for the "big" and "small" systems as in Fig. 4.32. The inverse of the slope is denoted by *DTL/DSMI*. Then draw a horizontal line at some value of *DTL/DSMI*!— say, at 5 — as shown in Fig. 4.32. This line intersects the two curves, as shown in Fig. 4.32, at two points: $TL = 65$ and $TL = 135$. Next, choose a value *higher* than $TL = 65$ — say, of $TL = 75$ — for the initial state of the "small" system, and similarly lower than $TL = 135$ — say, $TL = 125$ for the

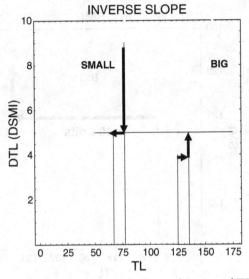

Fig. 4.32   The inverse slope as a function of *TL* for the "small" and the "big" systems.

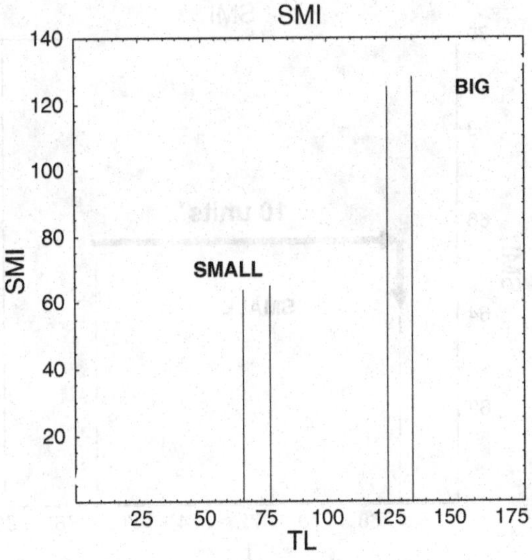

Fig. 4.33   The *SMI* versus *TL* for the "small" and "big" systems.

initial state of the "big" system. If you allow the combined systems to reach a new equilibrium state, both of the systems must reach an equilibrium state having the *same inverse slope*. Look at Fig. 4.32 and note that the *decrement* in the value of the inverse slope for the "small" system is much larger than the *increment* of the inverse slope of the "big" system. Now, go back to the *SMI* versus *TL* curve in Fig. 4.33 and its two amplifications in Fig. 4.34. You will see that the *increment* in the *SMI* of the "big" system is about twice the *decrement* of *SMI* of the "small" system. Therefore, the *net* change of *SMI* in this process is *positive*. You can repeat the same argument with the $\log_2 W$ versus *TL* curves, and you will find out that the *net* change in $\log_2 W$ is also *positive*. This means that 10 units of string length will be transferred *spontaneously* from the "small" to the "big" system. The reason is, of course, that the criterion for a spontaneous flow of string is always from the system with smaller inverse slope (higher temperature) to the larger inverse slope (lower temperature). See also Chapters 6 and 7.

6. Calculation of the Boltzmann distribution for the case of total length 9 (see Section 4.2). In calculating the Boltzmann distribution for the game $NM = 9$ and

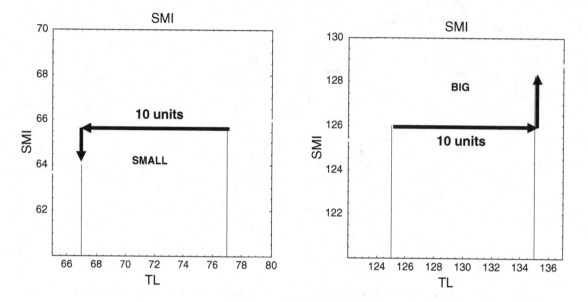

Fig. 4.34   Two amplifications of sections from Fig. 4.33.

$NC = 4$, we first solved the problem of maximizing the function

$$SMI = -\sum_{i=1}^{4} p_i \log p_i$$

Subject to the two conditions

$$\sum_{i=1}^{4} p_i = 1 \quad \text{and} \quad \sum_{i=1}^{4} (i-1)p_i = TL$$

Having found that solution in the form

$$p_i = Cx^i$$

The average number of marbles in each cell is obtained by multiplying p_i by 9.

The one we chose for the game in Section 4.2 was with $TL = 9$. This was chosen because it has the nice typical curve of the Boltzmann distribution.

For $TL = 13.5$, we recover the uniform distribution, which can be viewed as a particular case of the Boltzmann distribution having equal probability $p_i = 0.25$, hence

$$9 \sum_{i=1}^{4} (i-1)p_i = 9 \times 0.25 \sum_{i=1}^{4} (i-1) = 13.5$$

That means that the choice of length 13.5 (or average length per marble 13.5/9), gives a Boltzmann distribution that is uniform; that is, $p_i = 1/4$, or on average one marble per 4 cells.

When TL is larger than 13.5, we get an "inverted" Boltzmann distribution. This case corresponds to "negative temperature" where there is an inversion of population of energy levels.

Notes for Chapter 5

1. Figure 5.22 shows the theoretical probability distribution

$$\Pr(n/N) = \left(\frac{1}{2}\right)^N \frac{N!}{n!(N-n)!}$$

for a system of $N = 10, 50, 100$ marbles distributed in two boxes. n is the number of marbles in one box. The curves are plotted as a function of n/N, so they all have a maximum of $n/N = 1/2$. Note how the curve becomes smoother as the number of marbles increases. In the limit of very large N, the curve becomes the "Normal" or the "Gaussian" distribution.

2. To the reader who already identifies TA with the average kinetic energy of the particles, which in turn is related to the temperature, keeping $TA = 0$ is equivalent to keeping the system at absolute zero. In such a system there is no "shaking" of the system from the "inside." See also Chapter 7.

3. The case $TA = 200$ corresponds to the uniform distribution on 11 cells for which the probability distribution is $p(i) = \frac{1}{11}$ for each cell. If we have 20 marbles, the

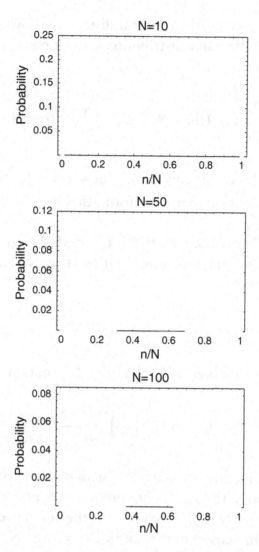

Fig. 5.22 The distribution Pr (n/N) in Note 1.

average number of marbles per cell will be $\frac{20}{11}$. The total cloth area for this case is

$$\frac{20}{11}\sum_{i=-5}^{5}i^2 = \frac{20}{11}(25+16+9+4+1+0+1+4+9+16+25) = \frac{110 \times 20}{11} = 200.$$

4. Although we shall not study the case of *TA* larger than 200, it is instructive to know that if we do increase *TA* beyond *TA* = 200 the curve of *SMI* starts to decrease with *TA*. In other words, the game becomes easier as we further increase *TA* beyond *TA* = 200. This region corresponds to negative temperatures and therefore is of no interest for real systems.

Notes for Chapter 6

1. Shannon's first theorem states that for a particle confined to a one-dimensional "box" of length L, the equilibrium probability distribution, $f_{eq}(x)$ is obtained by maximizing the *SMI*.

$$SMI = - \int_0^L f(x) \log f(x) \log dx \tag{1}$$

Subject to the normalization condition

$$\int_0^L f(x) dx = 1 \tag{2}$$

The result is (for details, see Ben-Naim (2008))

$$f_{eq}(x) = \frac{1}{L} \tag{3}$$

That is, $f_{eq}(x) dx$ is the probability of finding the particle between $x + dx$. The value of *SMI* for the equilibrium distribution is

$$SMI_{eq} = - \int_0^L f_{eq}(x) \log f_{eq}(x) dx = \log L \tag{4}$$

Assuming that the probability distributions along the three axes are independent, we write for the three dimensional case:

$$f_{eq}(x, y, z) = f_{eq}(x) f_{eq}(y) f_{eq}(z) = \frac{1}{L} \times \frac{1}{L} \times \frac{1}{L} = \frac{1}{V} \tag{5}$$

where we assumed that the particle is in a cube of edge L; that is, volume $V = L^3$.

In Chapter 3 we used the discrete version of this theorem; that is, we have n cells in one dimensional system and we maximize

$$SMI = -\sum_{i=1}^{n} p_i \log p_i \tag{6}$$

Subject to the condition that

$$\sum_{i=1}^{n} p_i = 1 \tag{7}$$

The result is

$$p_i = \frac{1}{n} \tag{8}$$

and the corresponding value of SMI is

$$SMI_{eq} = \log n \tag{9}$$

2. The SMI per particle is calculated by

$$SMI = -\sum_{i=1}^{n} p_i \log p_i \tag{1}$$

where $p_i = NM/NC$, which is the average density per cell.
The multiplicity is calculated by

$$W = \frac{NM!}{\prod_{i=1}^{NC} (p_i NM)!} \tag{2}$$

For large values of NM we can use the Stirling approximation to obtain the relationship

$$W \approx 2^{NM \times SMI} \tag{3}$$

Or equivalently

$$NM \times SMI \approx \log_2 W \tag{4}$$

For more details see Ben-Naim (2008).

3. The probability is related to W simply by

$$PR(NM_1, \ldots, NM_{NC}) = \left(\frac{1}{NC}\right)^{NM} W(NM_1, \ldots, NM_{NC})$$

$(NC)^{NM}$ is the total number of possible *specific* configurations. Therefore, $(NC)^{-NM}$ is the probability of finding one specific configuration of the system. This is a very small number when NM is large. $W(NM_1, \ldots, NM_{NC})$ is the number of specific configurations corresponding to the dim-arrangement $NM_1, \ldots, NM_{NC}$; that is, NM_1 marbles in cell 1, NM_2 in cell 2, and so on, NM_i is the number of marbles in cell i at the equilibrium state.

4. There is a subtle mathematical point that should be mentioned. As we have seen in Note 3, the probability of a *single* dim-state is always very small, even for the state having maximal value of W. However, the thermodynamic equilibrium state is not a single dim-state but a group of states at or near the state of maximum W. The sum of the probabilities of this group of states is nearly one. For more details, see Ben-Naim (2008).

5. Shannon's second theorem. The Boltzmann distribution is obtained by maximizing the *SMI*

$$SMI = -\int_0^\infty f(x) \log f(x) dx \tag{1}$$

Subjected to two constraints

$$\int_0^\infty f(x) dx = 1, \quad \int_0^\infty x f(x) dx = a, \quad \text{with } a > 0 \tag{2}$$

The function $f(x)$ that maximizes H, subjected to the two constraints (2) is

$$f_{eq}(x) = \frac{1}{a} \exp(-x/a) \tag{3}$$

The value of *SMI* for the distribution (3) is

$$SMI_{eq} = -\int_0^\infty f_{eq}(x) \log f_{eq}(x) dx = \log(ae) \tag{4}$$

The most common form of the Boltzmann distribution is

$$f(\varepsilon) = A \exp[-\varepsilon/k_B T] \tag{5}$$

where ε is the energy level, k_B the Boltzmann constant, and T the absolute temperature. A is a constant that is determined by the normalization condition.

A specific application of the Boltzmann distribution is the barometric distribution:

$$\rho(h) = \rho(h_o) \exp\left[-\frac{mg(h - h_o)}{k_B T}\right] \tag{6}$$

where $\rho(h)$ is the density of particles at a height h (relative to h_o), g is the gravitational acceleration, and m is the mass of particles. If we can assume that the gas is an ideal gas, then the pressure is given by

$$P(h) = \rho(h)k_B T \tag{7}$$

From (6) one can obtain the pressure distribution in a vertical column of air for a fixed temperature T.

6. W_{eq} is the number of specific states corresponding to the equilibrium dim-state (here the dim-state having the maximal value of W for a given system and a given TL). Denote by $W(TL)$ all possible specific states for a given value of TL, and by $W(TOTAL)$ all possible specific states of a system (with the fixed NM and NC, but any TL). Then, $W_{eq}/W(TOTAL)$ is the probability of finding the equilibrium dim-state and a specific value of TL. Therefore, the conditional probability of finding the equilibrium dim-state, given a fixed value of TL, is

$$PR(\textit{of the equilibrium state/given } TL) = \frac{W_{eq}}{W(TOTAL)} \frac{W(TOTAL)}{W(TL)} = \frac{W_{eq}}{W(TL)}$$

7. The Boltzmann distribution is usually written in the form (see Note 5)

$$f(\varepsilon) = A \exp[-\varepsilon/k_B T] \tag{1}$$

If all the energy of the system is kinetic energy, in one dimension, then one can simply write $\varepsilon = \frac{mv_x^2}{2}$ where m is the mass of each particle and v_x^2 is the velocity of the particles in the x direction. Substituting ε into (1) we get the typical exponent $\exp[-\frac{mv_x^2}{2k_B T}]$. To get the *MB* distribution we need to extend the range of variation of the velocities from $-\infty$ to $+\infty$, and normalize to obtain the *MB* distribution. See also Note 8.

8. Shannon's third theorem. In one dimension, say the x axis, the kinetic energy of a particle of mass m is given by

$$\varepsilon_x = \frac{mv_x^2}{2} \tag{1}$$

Substituting ε_x in the Boltzmann distribution Eq. (5), in Note 5, we get

$$f(v_x) = A\exp\left[\frac{-mv_x^2}{2k_BT}\right] \tag{2}$$

Note that this distribution is symmetrical with respect to zero; that is, the probability density of obtaining a velocity v_x is equal to that of obtaining a velocity $-v_x$. The normalization constant A is obtained by requiring that

$$\int_{-\infty}^{\infty} f(v_x)dv_x = 1 \tag{3}$$

and the result is

$$f_{eq}(v_x) = \sqrt{\frac{m}{2\pi k_BT}}\exp\left[\frac{-mv_x^2}{2k_BT}\right] \tag{4}$$

One can also obtain this distribution by maximizing the *SMI*, subjected to the two conditions:

$$\int_{-\infty}^{\infty} f(v_x)dv_x = 1, \quad \int_{-\infty}^{\infty} v_x^2 f(v_x)dv_x = \sigma^2 \tag{5}$$

The value of *SMI* for the equilibrium distribution of velocities in one dimension is

$$SMI_{eq} = \frac{1}{2}\log\left(2\pi eT/m\right) \tag{6}$$

To obtain the distribution of *speeds*; that is, the positive quantity defined by

$$v = \sqrt{v_x^2 + v_y^2 + v_z^2} \tag{7}$$

One assumes that the distribution of velocities v_x, v_y and v_z are independent, hence

$$f_{eq}(v_x, v_y, v_z) = f_{eq}(v_x)f_{eq}(v_y)f_{eq}(v_z) = \left(\frac{m}{2\pi k_B T}\right)^{3/2} \exp\left[\frac{-m}{2k_B T}(v_x^2 + v_y^2 + v_z^2)\right]$$

$$= \left(\frac{m}{2\pi k_B T}\right)^{3/2} \exp\left[\frac{-mv^2}{2k_B T}\right] \tag{8}$$

The distribution of speed (v) is obtained by transforming to spherical polar coordinates, and integrating over the angles. The result is

$$f_{eq}(v) = \left(\frac{m}{2\pi k_B T}\right)^{3/2} 4\pi v^2 \exp\left[\frac{-mv^2}{2k_B T}\right] \tag{9}$$

See illustration in Fig. 6.2.

9. We must be careful here. The initial state of the combined system is *after* we have lifted the constraint of constant TL for each system, and allowed the free exchange of string between the two systems.

10. In real life, we need to add that the system under consideration is *isolated*; that is, there are no interactions between the system and the environment. In all our experiments, the isolation of the entire system (whether one, two, or more games) was made implicitly.

Notes for Chapter 7

1. For more details see Ben-Naim (2008).
2. In thermodynamics, the result is $\Delta S = k_B \ln(2V/V) = k_B \ln 2$ where k_B is the Boltzmann constant.
3. See, for example, Lindley (1965). Mathematically, the probability of an event A should be written as $p(A/\Omega)$ where Ω is the sample space. Usually this notation is replaced with $P(A)$, but it always implies the condition "given Ω". For more details, see Ben-Naim (2008).
4. Gibbs (1906).

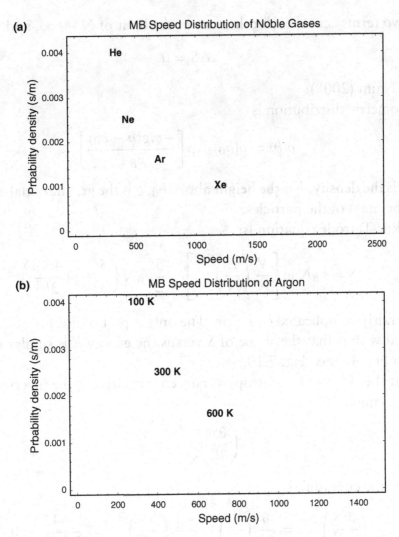

Fig. 6.2 (a) The speed distribution of noble gases at $T = 298K$. (b) The speed distribution of argon at three different temperatures.

5. Using the partition function of an ideal gas, it is easy to show that in process 7.6b there are two contributions to the entropy change

$$\Delta S = 2N \log 2 - \log \frac{(2N)!}{(N!)^2}$$

These two terms cancel each other out in the limit of $N \to \infty$, and we get

$$\Delta S = 0$$

6. See Ben-Naim (2008).
7. The Barometric distribution is

$$\rho(h) = \rho(h_0) \exp\left[\frac{-mg(h - h_0)}{k_B T}\right]$$

where ρ is the density, h is the height above h_0, g is the gravitational acceleration, and m the mass of the particles.
8. The Sackur-Tetrode equation is:

$$S = k_B N \ln\left[\frac{V}{N}\left(\frac{E}{N}\right)^{3/2}\right] + \frac{3}{2}k_B N\left(\frac{5}{3} + \ln\frac{4\pi m}{3h^2}\right)$$

This is a fairly complicated equation. The only aspect of this formula that we are concerned with is that the shape of S versus the energy E is similar to the curve shown in Fig. 4.20 or Fig. 7.10.
9. Note that the slope of the entropy versus energy curve is the inverse of the temperature, namely:

$$\left(\frac{\partial S}{\partial E}\right)_{N,V} = \frac{1}{T}$$

Therefore the curvature is

$$\left(\frac{\partial^2 S}{\partial E^2}\right)_{N,V} = \frac{\partial}{\partial E}\left(\frac{1}{T}\right) = \frac{-1}{T^2}\left(\frac{\partial T}{\partial E}\right)_{N,V} = \frac{-1}{T^2 C_V} < 0$$

Since C_V must always be positive, the second derivative of S with respect to E is always negative.
10. All we say here about the one-dimensional case also applies to the three-dimensional case, except that the latter is a little more complicated.
11. The kinetic energy of a moving body in $mv^2/2$, where m is the mass and v the speed of the body.

12. What we do is simply transformation of variables. More specifically, if the Boltzmann distribution is written as $P_B(\varepsilon) = C2^{-\lambda\varepsilon}$, where C and λ are constants, and if $\varepsilon = N\frac{mv^2}{2}$, then the new distribution is $P_{MB}(v) = C2^{(\frac{-mv^2}{2}\lambda)}$.

13. In thermodynamics, we have the following equation for the change in entropy of an ideal gas, between two states, 1 and 2:

$$\Delta S = S_2 - S_1 = C_V \ln \frac{V_2}{V_1} + R \ln \frac{T_2}{T_1}$$

where C_V is the heat capacity at constant volume and R the gas constant. It is relatively easy to explain the entropy change associated with the *volume* change in the system. This change involves redistribution of particles in their locations. It is not straightforward to explain the entropy change associated with the temperature changes in the system. To do the latter we need to understand the change in velocity distributions associated with the changes in the temperature.

14. Here we use the classical description of the system. The quantum mechanical description is more abstract. Yet, somewhat paradoxically, the formulation of the Second Law and entropy are simpler in the quantum mechanical description of the system.

15. In information theory, a continuous variable would entail an infinite amount of information to describe. For this reason, it is advisable to divide both the locational and the velocity range into discrete cells (see Fig. 7.12).

16. The relationship between the super-probability and the entropy is shown in Fig. 7.15. For more details see Ben-Naim (2008).

$$PR(\{p_1,...,p_n\}) = \left(\frac{1}{n}\right)^N \frac{2^{N \times S(\{p_1,...,p_n\})}}{\sqrt{2\pi N^{(n-1)} \prod\limits_{i=1}^{n} p_i}}$$

Fig. 7.15 The mathematical relationship between the super probability PR and the entropy S at equilibrium.

Epilogue

As I stated in the Preface, my aim in writing this book was to show how easy it is to "discover" and to understand entropy and the Second Law. To achieve that goal, I have developed a methodology for guiding you, the reader, on how to do "research," how to analyze the results of an experiment, and how to make sense of these results.

On several occasions while writing the book, I felt an eerie sense of being affected by the same methods I was using to guide you in understanding the Second Law. The very methods I developed to show the reader how to discover new things turned out to reflect back on me — unraveling to me many new things that I did not know or that I did not even plan to write about.

As an example, I did not plan to write anything about temperature. While I had a clear idea of how to introduce the analog of entropy in the world of marbles, I had no idea even of how to *define* temperature in such games. I was satisfied with the fact that both entropy and the Second Law have their analogs in the games of marbles. That was more than enough to fulfill my goals in explaining the Second Law. Yet, while doing the experiments in Chapters 4 and 5 on the spontaneous transfer of string and cloth, I thought that some of the experiments might lead to wrong conclusions. That is how I finally decided to add Sections 4.7 and 4.8, and some further discussion of the analog of the temperature in the games of marbles.

Remember that after doing the experiment in Section 4.7, I asked you, the reader, to explain why the flow of string would stop. If you read carefully what I have written there, you will see that I first accepted your answer, but soon after that I realized that my endorsement of your answer could be misleading. As it was "natural" to conclude from that particular experiment that energy will flow from the high energy system to the low energy system, this conclusion is correct only if the two systems are identical except for their energies. However, in general, that conclusion is incorrect. Therefore, I had to find out how to introduce the analog of temperature in the games, and to show that energy (or string or cloth) flows from higher to lower temperature. That was not easy to find, but eventually I added Section 4.8 as well as the "most challenging exercise" in Section 4.8.

This is perhaps one of the most surprising aspects of writing a book of this kind — you try to teach the reader how to search for a "discovery", and then you end up discovering something you never knew before.

The other thing that I have "discovered" while writing the book was the intimate relationship between the question of *what* entropy is, and *why* it always changes in one direction only. These two questions are, in general, very different. One is determined by the *SMI* of a thermodynamic system, while the other is governed by the conditional probability of the transition from the initial to the final states. In general, the *SMI* and the probability of the state seem to be unrelated concepts and yet they are intimately intertwined. Both entropy (or the *SMI*) and the probability of the state of the system (which I referred to as the super probability) are defined *on* a probability distribution. This in itself does not guarantee a relationship between the two. One can define two functions — say, $f = f(x)$ and $g = g(x)$ — which in general are not related to each other (this is especially true when x stands for many variables $(x_1, \ldots, x_n)$). However, while doing the experiments it became clear to me that $\log W$ or $\log \Pr$ is related to *SMI*. In fact, one can prove that in the limit of very large systems, $\log \Pr$ is proportional to N times the *SMI*. The formula, along with its symbolic significance are shown in Figs. 7.13 and 7.14. It was much later that I asked Alex Vaisman to draw the symbolic Fig. 7.15, a version of which was chosen for the cover design of the book.

In summary, the writing of this book turned out to be much more difficult than I had initially expected. It is not an easy task to translate into simple language and explain to a layperson concepts that are not only difficult but sometimes mysterious even to scientists.

While writing this book, I had many email exchanges with numerous friends and colleagues, as well as people who have written to me and commented on my previous book.

Thanks for all the comments. I have tried my best to learn from previous errors to improve the style and the clarity of my writing. I hope I have achieved that goal. I will appreciate any comment from any reader of this book.

Marc Marschark drew my attention to a quotation from Einstein which he quoted in his book. I will reproduce it here:

> Anyone who has ever tried to present a rather abstract scientific subject in a popular manner knows the great difficulties of such an attempt. Either he succeeds in being intelligible by concealing the core of the problem and by offering the reader only superficial aspects or vague illusions, thus deceiving the reader by arousing in him the deceptive illusion of comprehension; or else he gives an expert account of the problem, but in such a fashion that the untrained reader is unable to follow the exposition and becomes discouraged from reading further. If these two categories are omitted from today's popular scientific literature, surprisingly little remains ... It is of great importance that the general public be given an opportunity to experience — consciously and intelligently — the efforts and results of scientific research. It is not sufficient that each result be taken up, elaborated, and applied by a few specialists in the field. (Quoted in Barnett (1957))

Postscript

After I finished writing this book, I heard the sad news of the passing away of my friend and colleague Daniel Amit. It reminded me of a short correspondence I had with Dani a few years ago. Sometime in 2006, I sent him the manuscript of my earlier book *Entropy Demystified*, for comment and review. He agreed to read and comment on the manuscript, although he wrote to me that he had had a stroke and that his health had been deteriorating. After reading the preface and the introduction, Dani commented that the book seemed to be challenging, but he added: "I personally do not believe that entropy could ever be demystified." I wrote back to ask him, "What if I succeed to demystify you about entropy; would you agree on my publishing our correspondence?"? He wrote back one word: "Agree." I never heard from him again until I read about the news of his passing away.

I myself was mystified by entropy for so many years until I learned of Shannon's theory through the works of Jaynes and Katz. At that point I was totally demystified about entropy. Instead, I became more mystified by the fact that entropy has been a mystery for so long ...

References and Suggested Reading

Attneave F. (1959) *Application of Information Theory to Psychology*. Henry Holt and Company, New York.

Barnett L. (1957) *The Universe and Dr Einstein*. Dover Publication, New York.

Bell CR. ed. (1979) *Uncertain Outcomes*. MTP Press Ltd, Lancaster.

Bendig AA. (1953) *J Exp Psychol* **46**: 345.

Ben-Naim A. (1987) *Am J Phys* **55**: 1105.

Ben-Naim A. (2007) *Entropy Demystified*. World Scientific Publishing, Singapore.

Ben-Naim A. (2008) *A Farewell to Entropy: Statistical Thermodynamics Based on Information*. World Scientific, Singapore.

Bruner JS, Olver RR, Greenfield PM. eds. (1966) *Studies in Cognitive Growth*. John Wiley and Sons, Inc. New York.

Callen HB. (1985) *Thermodynamics and an Introduction to Thermostatics*, 2nd ed. John Wiley and Sons.

Cooper LN. (1968) *An Introduction to the Meaning and Structure of Physics*. Harper and Row, New York.

Courage ML. (1989) *Child Dev* **60**: 877.

Denbigh KG. (1966) *Principles of Chemical Equilibrium*. Cambridge University Press, London.

Denney NW. (1980) "The effect of the manipulation of a referral, non-cognitive variables on the problem-solving performance for the elderly," *Hum Dev* **23**: 268.

Denney NW. (1985) "A review of life span research with the twenty questions task: A study of problem-solving ability," *Int J Aging Hum Dev* **21**: 161.

Denney NW, Denney DR, Ziobrowski M. (1973) "Alteration in the information-processing strategies of young children following observation of adult models," *Dev Psychol* **8**: 202.

Denney NW, Turner MC. (1979) "Facilitating cognitive performance in children: A comparison of strategy modeling and strategy modeling with overt self-verbalization," *J Exp Child Psychol* **28**: 119.

Drumm P, Jackson DW, Magley V. (1995) "The primacy of superordinate-level category questions in the game of 20 questions," *Percept Mot Skills* **81**: 271.

Falk R. (1983) *Proceedings of the First International Conference on Teaching Statistics*, Vol. II. University of Sheffield, p. 714.

Falk R, Falk R, Levin I. (1980) "A potential for learning probability in young children," *Educ Studies Math* **11**: 181.

Falk R, Wilkening F. (1998) *Dev Psychol* **34**: 1340.

Fast JD. (1962) *Entropy, the Significance of the Concept of Entropy and its Applications in Science and Technology*. Philips Technical Library.

Feller W. (1950) *An Introduction to Probability Theory and its Application*. John Wiley and Sons, New York.

Fischbein E. (1975) *The Intuitive Sources of Probabilistic Thinking in Children*. D. Reidel Publishing Company, Boston.

Gibbs JW. (1906) *Scientific Papers, Thermodynamics*, Vol. 1. Longmans Green, New York.

Jaynes ET. (1957) Information theory and statistical mechanics. *Phys Rev* 106: 620.

Katz A. (1967) *Principles of Statistical Mechanics, The Information Theory Approach*. WH Freeman and Co., San Francisco.

Kusche CA, Greenberg MT. (1983) Evaluative understanding and role-taking ability: A comparison of deaf and hearing children. *Child Dev* 54: 141.

Kusche CA, Greenberg MT, Garfield TS. (1983) Nonverbal intelligence and verbal achievement in deaf adolescents: An examination of heredity and environment. *Am Ann Deaf* 128: 458.

Leake L, Burrell P, Fischbein HD. trans. (1975) *The Origin of Chance in Children*. Norton, New York.

Lewis GN. (1930) The symmetry of time in physics. *Science* 71: 569.

Lindley DV. (1965) *Introduction to Probability and Statistics*. Cambridge University Press, Cambridge.

Luckner JL, McNeill JH. (1994) "Performance of a group of deaf and hard-of-hearing students and a comparison group of hearing students on a series of problem-solving tasks," *Am Ann Deaf* 139: 371.

Marschark M, Everhart VS. (1999) *Problem-Solving by Deaf and Hearing Students: Twenty Questions in Deafness and Education International*. Whuzz Publishers Ltd.

Miller GA, Frick FC. (1949) Statistical behavioristics and sequence of responses. *Psychol Rev* 56: 311.

Mosher FA, Hornsby JR. (1966) in Chapter 4, "On asking questions." In: Bruner JS, Olver RR, Greenfield PM (eds), *Studies in Cognitive Growth*. John Wiley and Sons, Inc, New York.

Mousley K, Kelly RR. (1998) Problem-solving strategies for teaching mathematics to deaf students. *Am Ann Deaf* 143: 325.

Olver RR, Hornsby JR. (1966) in Chapter 3, "Equivalence." In: Bruner JS, Olver RR, Greenfield PM (eds), *Studies in Cognitive Growth*. John Wiley and Sons, Inc, New York.

Papoulis A. (1965) *Probability, Random Variables and Stochastic Processes*. McGraw Hill, New York.

Piaget J, Inhelder B. (1951) *La Genese de l'Idee de Hazard Chez l' Enfant*. PUF, Paris.

Shannon CE. (1948) Mathematical theory of communication. *Bell Syst Tech J* 27: 379, 623.

Siegler RS. (1977) *Child Dev* 48: 395.

Tribus M, McIrvine EC. (1971) Entropy and information. *Sci Am* 225: 179.

Wiener N. (1948) *Cybernetics*. John Wiley, New York.

Index